Managing the Digital Transformation

This book provides the key technologies involved in an organization's digital transformation. It offers a deep understanding of the key technologies (Blockchain, AI, Big Data, IoT, etc.) involved and details the impact, the decision-making process, and the interplay between technologies, business models, and operations.

Managing the Digital Transformation: Aligning Technologies, Business Models, and Operations provides frameworks and models to support digital transformation projects. The book presents the importance of digital transformation as a resilience approach to the operations processes and business models. It covers the essential elements integrating the technology, the organizations, the operations, and supply chain management used to move toward digital transformation. Concepts and mini-case studies are included to provide a deeper understanding of digital transformation projects with a holistic view. The book also examines the role that digital transformation plays with consideration of inter-organizational and intra-organizational capabilities, along with the role of digital culture, the worker's skills, business models, reconfiguration, as well as an operations optimization angle.

Practitioners, consultants, governments, managers, scholars, and anyone interested in digital transformation will find the contents of this book very useful.

Emerging Operations Research Methodologies and Applications

Series Editors:
Natarajan Gautam
Texas A&M, College Station, USA

A. Ravi Ravindran
*The Pennsylvania State University,
University Park, USA*

Multiple Objective Analytics for Criminal Justice Systems
Gerald W. Evans

Design and Analysis of Closed-Loop Supply Chain Networks
Subramanian Pazhani

Social Media Analytics and Practical Applications
The Change to the Competition Landscape
Subodha Kumar and Liangfei Qiu

Analysis of Fork-Join Systems
Network of Queues with Precedence Constraints
Samyukta Sethuraman

Managing the Digital Transformation
Aligning Technologies, Business Models, and Operations
Maciel M. Queiroz and Samuel Fosso Wamba

For more information about this series, please visit: https://www.routledge.com/Emerging-Operations-Research-Methodologies-and-Applications/book-series/CRCEORMA

Managing the Digital Transformation
Aligning Technologies, Business Models, and Operations

Maciel M. Queiroz and Samuel Fosso Wamba

CRC Press
Taylor & Francis Group
Boca Raton London New York

CRC Press is an imprint of the
Taylor & Francis Group, an **informa** business

First edition published 2023
by CRC Press
6000 Broken Sound Parkway NW, Suite 300, Boca Raton, FL 33487-2742

and by CRC Press
4 Park Square, Milton Park, Abingdon, Oxon, OX14 4RN

CRC Press is an imprint of Taylor & Francis Group, LLC

ISBN: 978-1-032-12850-4 (hbk)
ISBN: 978-1-032-12852-8 (pbk)
ISBN: 978-1-003-22646-8 (ebk)

DOI: 10.1201/9781003226468

Typeset in Times
by codeMantra

Contents

Preface

In recent years, the digital transformation phenomenon with the unprecedented advance of information and communication technologies (ICTs), cutting-edge and highly disruptive technologies, has emerged and captured the imagination of scholars, practitioners, managers, government, and decision-makers. More specifically, different technologies such as artificial intelligence (AI), big data analytics (BDA), internet of things (IoT), blockchain, cloud computing, and 3D printing, among others, are remodeling business models from representative and traditional industries.

There is a race between the organizations and their supply chains concerning the digitalization of their business models. In addition, scholars are trying to understand the dynamics of the adoption, implementation, and diffusion of key technologies toward digital transformation. However, the adoption of the technology per se does not ensure the success of a digital transformation project. In this outlook, many digital transformation projects fail. A holistic approach can minimize the high rates of unsuccessful digital transformation initiatives.

In this context, the extant literature did not yet provide a robust and integrated view considering technology, organization, and people to support the operations and supply chain, taking into account the digital transformation perspective. It has focused primarily on technology capabilities. Thus, our book fills this practical and theoretical gap by exploring essential topics covering the digital transformation at the organization and supply chain levels. For example, we examine the role of the digital transformation considering inter-organizational and intra-organizational capabilities, the role of digital culture, the worker's skills, the business model reconfiguration, and the operations optimization angle. Our book also brings to the arena the importance of digital transformation as a resilience approach to operations processes and business models.

Considering this background, we provide essential elements integrating the technology, organizations, and people in the operations and supply chain management toward the digital transformation. From this perspective, this book offers easy concepts, short-cases, frameworks, and models to leverage the ideas for all interested in gaining a deeper understanding of digital transformation projects in a holistic view. Thus, we have some examples of critical questions that are covered throughout the book:

- Which are the roles and foundations of digital transformation?
- What are the key technologies that support digital transformation at different stages of diffusion?
- How could organizations integrate digital capabilities considering inter-organizational and intra-organizational perspectives?
- Which are the barriers and enablers in digital transformation projects?
- What is the role of data-driven and organization's digital culture?
- What are people's roles, challenges, and opportunities in a digitalized world?
- What are the worker's skills required for the digital transformation age?
- How performance improvement and competitive advantage could be leveraged by digital transformation?
- What is the role of operations research in the digital transformation age?
- What are the main benefits of integrating operations management, operations research, and digital transformation in supply chains?
- How can digital transformation enabled by operations research minimize uncertainty in the supply chain?

This book presents singular value to practitioners from all types of industries. Also, it intends to be one of the main sources for organizations in different stages of digital transformation projects. Furthermore, our book can be used by business schools and universities in different courses and levels, including MBA and PhD degrees and graduate, undergraduate, and executive courses.

Acknowledgments

This book results from an intensive and excellent collaboration of scholars, industry practitioners, and the CRC Press's staff. We especially thank all the reviewers who provided exceptional feedback and suggestions to the authors. In addition, we would like to give special thanks to Cindy Carelli, Executive Editor at CRC Press, and Erin Harris, Senior Editorial Assistant at CRC Press. Also, we extend our thanks to all staff of the CRC Press—Taylor & Francis Group for their outstanding assistance during the whole process of this project. Finally, we thank our families and friends for their invaluable support and motivation for the accomplishment of this book.

Maciel M. Queiroz
Samuel Fosso Wamba

About the Authors

Dr Maciel M. Queiroz is an Associate Professor and Researcher of Operations and Supply Chain Management at FGV EAESP, Brazil. He earned his PhD in Naval Architecture and Ocean Engineering at the University of Sao Paulo, Brazil. Currently, Maciel is Associate Editor in the International Journal of Logistics Management, RAUSP Management Journal, and Editorial Review Board of the International Journal of Information Management. His current research focuses on digital transformation, including digital supply chain capabilities, Industry 4.0, AI, blockchain, big data, and IoT. He has published papers in top-tier international journals and conferences. Besides, Dr Maciel has been serving as a Guest Editor for leading journals, including the *International Journal of Operations & Production Management*, the *Journal of Business Logistics*, the *International Journal of Production Research*, *Production Planning and Control*, *Annals of Operations Research*, the *International Journal of Information Management*, the *Journal of Enterprise Information Management*, *Industrial Management and Data Systems,* etc.

Dr Samuel Fosso Wamba is the Associated Dean of Research at TBS Education, France. He is also a Distinguished Visiting Professor at The University of Johannesburg, South Africa, and at the UCSI Graduate Business School, UCSI University, Malaysia. He earned his PhD in industrial engineering from the Polytechnic School of Montreal, Canada. His current research focuses on the business value of information technology, inter-organizational systems adoption, use and impacts, supply chain management, electronic commerce, blockchain, artificial intelligence for business, social media, business analytics, big data, and open data. He leads the Center of Excellence in Artificial Intelligence & Business Analytics at TBS Education. He is among the 2% of the most influential scholars globally based on the Mendeley database that includes 100,000 top scientists for 2020 and 2021. He ranks in ClarivateTMs 1% most cited scholars in the world for 2020 and 2021 and in CDO Magazine's Leading Academic Data Leaders 2021. Based on the Research.com 2021 ranking, he is France's third top business and management scientist.

Foundations of the Digital Transformation

1

1.1 INTRODUCTION: THE ROLE OF DIGITAL TRANSFORMATION

In recent years, the digital transformation phenomenon has captured the attention and imagination of practitioners, scholars, and society (Laster, 2021; Queiroz et al., 2021). Digital transformation is a broad concept that is related not only to digital technologies but also involves a set of different types of resources (including humans and organizations) that should be integrated in a strategic and harmonious way (Tabrizi et al., 2019). There is some confusion regarding several of the key terms of digital transformation.

This chapter does not aim to provide a literature review on the main terms. Rather, it outlines and connects the main differences between digitization, digitalization, and digital transformation:

- **Digitization** refers to the conversion of an analog process, which essentially includes physical processes/activities that are converted to a digital business model, usually by adopting information and communication technologies (ICTs) (Legner et al., 2017; Queiroz et al., 2021).

- **Digitalization** is the next step. It refers to the process after such technologies are adopted and implemented and the effect on an organization's business models, as well as on government and society (Legner et al., 2017; Queiroz et al., 2021). As such, it is clear that the digitization process is focused on technology adoption and its application in the physical processes and activities, while digitalization is related to the impacts (positive/negative) of the transformation on the company and its stakeholders (Queiroz et al., 2021).
- **Digital transformation** is the third stage and refers to an organization's continuous process of creating value for society (i.e., customers, suppliers, government, other organizations, etc.) by integrating cutting-edge technologies and skilled people (Legner et al., 2017; Queiroz et al., 2021).

Figure 1.1 highlights the digital transformation as the main dimension, which is supported by the digitization and digitalization concepts. Accordingly, there can be no digital transformation without the digitization of physical activities or their adoption in the business processes. It is also important to note that digital transformation is a continuous process that has no end. After any analog activity is converted into a digital version, the business process

FIGURE 1.1   Digitization, digitalization, and digital transformation. Adapted from Queiroz et al. (2021).

tends to be optimized, which in turn usually positively impacts the organization's value creation (e.g., greater agility, responsiveness, and visibility
among customers).

1.2 TECHNOLOGIES, ORGANIZATIONS, AND PEOPLE: THE ESSENTIAL VIEW OF THE DIGITAL TRANSFORMATION

Unfortunately, there is some confusion in academia and industry related to
digital transformation. However, the term "digital" leads many companies—
wrongly—to merely adopt technologies without having a strategic plan for
the future that considers other essential aspects.

In this book, we call readers' attention to the intersection that defines
and is key to the digital transformation: technologies, organizations, and
people (TOP). Figure 1.2, adapted from Bordeleau et al. (2021) shows how
these three concepts intersect. In order to facilitate the presentation of the
digital transformation phenomenon throughout this book, we will craft our

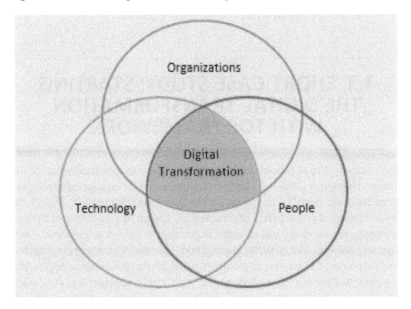

FIGURE 1.2 TOP framework for the digital transformation. Adapted from
Bordeleau et al. (2021).

argument following the TOP framework. We refer to each part of the TOP as a pillar. Accordingly, each pillar has the same importance and influence on digital transformation. In this chapter, we present an overview of each pillar, which is explored in greater depth in the following chapters.

- **Technology:** This is undoubtedly the best-known pillar of digital transformation. Usually, it is focused on or associated with cutting-edge technologies, also known as emerging technologies (see Chapter 2 for more details). However, established (traditional) technologies are also important to an organization's digital transformation strategies.
- **Organizations:** From the digital transformation standpoint, organizations play a critical role in providing adequate resources and capabilities (i.e., infrastructure, strategic planning, leadership, top management support, data-driven cultures, etc.) to the firm's digital transformation process (see Chapter 3 for more details).
- **People:** A large number of organizations do not pay adequate attention to this pillar of their digital transformation strategies. People are the cornerstone of any type of digital transformation project. If companies are to adopt and implement digital technologies and remodel their processes to create value for stakeholders, people are essential (see Chapter 3 for more details).

1.3 SHORT CASE STUDY: STARTING THE DIGITAL TRANSFORMATION WITH TOP FRAMEWORK

The company Top is a multinational organization that produces chips for the automotive industry. During a recent meeting, a middle manager shared some ideas about the need for an urgent digital transformation project. One of the senior managers immediately interrupted his speech, noting that over the last two years, the company had invested millions of dollars in acquiring cutting-edge technologies like artificial intelligence. This senior manager emphasized that despite the substantial amounts invested in disruptive technologies, the company had not maximized its profits. As a result, management concluded that the company's digital transformation project had failed. Based on this scenario, list the main elements that this company employed in its attempts at digital transformation. Why was this project considered unsuccessful? How can the TOP framework be used in this scenario?

REFERENCES

Bordeleau, F.-È., Santa-Eulalia, L. A., & Mosconi, E. (2021). Digital transformation framework: Creating sensing, smart, sustainable and social (S^4) organisations. *HICSS*. https://doi.org/10.24251/HICSS.2021.559

Laster, D. (2021). *Why the Era of Digital Transformation Is Important for Companies of All Sizes.* Forbes. https://www.forbes.com/sites/forbesbusiness-council/2021/09/23/why-the-era-of-digital-transformation-is-important-for-companies-of-all-sizes/?sh=530f25737a12

Legner, C., Eymann, T., Hess, T., Matt, C., Böhmann, T., Drews, P., Mädche, A., Urbach, N., & Ahlemann, F. (2017). Digitalization: Opportunity and challenge for the business and information systems engineering community. *Business & Information Systems Engineering*, 59(4), 301–308. https://doi.org/10.1007/s12599-017-0484-2

Queiroz, M. M., Pereira, S. C. F., Telles, R., & Machado, M. C. (2021). Industry 4.0 and digital supply chain capabilities. *Benchmarking: An International Journal*, 28(5), 1761–1782. https://doi.org/10.1108/BIJ-12-2018-0435

Tabrizi, B., Lam, E., Girard, K., & Irvin, V. (2019). *Digital Transformation Is Not About Technology.* Harvard Business Review. https://hbr.org/2019/03/digital-transformation-is-not-about-technology

Key Technologies for Digital Transformation

2

2.1 INDUSTRY 4.0 AS A TRIGGER TO DIGITAL TRANSFORMATION

The Fourth Industrial Revolution (Schwab, 2016; Caiado et al., 2021), also known as Industry 4.0 or Advanced Manufacturing in some countries, is a concept that emerged in 2011 during the Hannover Messe 2011 industrial trade fair. It is supported mainly by emerging technologies (Queiroz et al., 2021), which are defined as highly disruptive technologies that are not yet fully validated and widespread. One of the main premises of Industry 4.0 is that such technologies allow autonomous communication between different types of equipment such as machines, products, and materials throughout the entire production system. This communication and interaction are not restricted to equipment (devices) but occur with factories, customers, and the entire supply chain.

Moreover, Industry 4.0 is strongly related to the digital transformation of organizations, generating connectivity between people, objects, and systems, thus enabling the exchange of data in real-time (Meindl et al., 2021). In this

DOI: 10.1201/9781003226468-2

context, it is clear that Industry 4.0 is already reshaping not just manufacturing but all types of industries and businesses. By using a combination of different types of technologies, a new generation of smart products and customers is emerging, requiring a complete transformation of organizations and their business models.

2.2 A BRIEF DESCRIPTION OF THE MOST COMMON DIGITAL TRANSFORMATION TECHNOLOGIES

This subsection shows some of the key technologies used in digital transformation projects. The objective is not to provide an exhaustive list of all the available technologies because in today's world, technologies are evolving at an unprecedented and exponential rate. Table 2.1 briefly describes these key technologies and provides several key references for readers who are interested in gaining a more in-depth understanding. It is important to note that the technologies are at different stages of maturity; examples include both cutting-edge technologies (e.g., blockchain, 3D printing, and big data analytics) and established technologies (e.g., EDI, RFID, GPS).

2.3 BENEFITS ENABLED BY THE TECHNOLOGIES OF THE DIGITAL TRANSFORMATION

The technologies mentioned in the previous sections have the power to remodel and improve the business processes of any type of organization. For instance, with IoT applications, manufacturing companies can collect data from their production systems and network and analyze them with BDA and AI tools. IoT enables a vast range of uses, such as developing prediction models to improve maintenance activities, improving performance in operations and production management through enhanced monitoring processes of the objects, and collecting feedback from customers.

With blockchain, companies can track the provenance of products, and consumers can monitor the journey of a particular product until its reaches their home. Blockchain can also decentralize business models and minimize

TABLE 2.1   Popular technologies of the digital transformation

KEY TECHNOLOGIES	BRIEF DESCRIPTION	REFERENCES
Internet of things (IoT)	Refers to the network of physical objects embedded with sensors, which can communicate with each other to exchange data with different types of devices, supported by mainly the internet.	(Queiroz et al., 2020)
Cyber-physical systems (CPS)	Refers to the integration of physical processes and networking, supported by information communication technologies (ICTs). Traditionally, CPS is used to control the physical object's flow in production systems, providing feedback through the process.	(J. Lee et al., 2017)
Big data analytics (BDA)	Refers to techniques used to capture, store, manage, and analyze large amounts of data sets whose size outperforms typical database software approaches. The data sets can originate from different sources (internet, organizations, GPS, etc.) and are analyzed by using descriptive, prescriptive, and predictive approaches.	(Fosso Wamba et al., 2020)
Cloud computing (CC)	Refers to a set of computing services over the internet. CC services can offer different types of services flexibly, cost less, and be optimized. Some of the popular services are data storage, software, servers, and business intelligence, among others.	(Charro & Schaefer, 2018)
Artificial intelligence (AI)	Refers to machines (computers, software, algorithms, etc.) that perform activities that require some human intelligence, that is, machines and devices with intelligence to perform tasks without or with human interaction.	(Fosso Wamba et al., 2021)
Simulation (SI)	Refers to the modeling of the real-world process, problem, or situation, which is performed by specific software to gain and provide a more in-depth understanding of the problem or phenomenon. SI is a traditional approach to testing different configurations of processes, projects and activity changes before implementation.	(Ivanov, 2020)

Virtual reality (VR)	Refers to a simulated three-dimensional environment by using an electronic device with a screen and sensors. VR applications provide a sensory experience by enabling the users to interact with the objects in a fictional manner. These applications can be used for a vast range of contexts, including corporate training, rehabilitation treatments, games, etc.	(Alzayat & Lee, 2021)
Augmented reality (AR)	Refers to applications for an interactive experience of the real world in which the users can interact with real objects through visual devices. AR has been used to leverage the experience of the user in a specific context. For instance, AR can enhance visualization in a warehouse for picking activities.	(Joerß et al., 2021)
3D printing/ Additive manufacturing (3DP/AM)	Refers to a process that produces three-dimensional objects created in a digital file. 3DP/AM supports rapid prototyping in the production process, consequently enabling decentralized and on-demand production.	(Beltagui et al., 2020)
Digital twins (DTw)	Refers to a representation (copy) of processes, physical objects, organizations or supply chains (SCs) with real-time updates. DTw enables firms and SCs to monitor in real-time the entire or part of their operations by detecting, preventing, predicting and optimizing the operations.	(Ivanov, 2021)
Blockchain technology (BT)	Refers to a distributed database (the ledger of a book) that permits peer-to-peer transactions with a high level of cryptography. BT is based on a decentralized model, with no central authority to validate the transactions. It can improve the agility and the costs of the transactions using an approach that allows traceability of any transaction.	(Wamba & Queiroz, 2022)
Quantum computing (QC)	Refers to the use of quantic mechanics logic to execute processes and tasks in a faster way when compared to other computing approaches. QC perform the processes by using superposition, interference, and entanglement, which are quantum state properties.	(Ajagekar & You, 2019)

(Continued)

TABLE 2.1   (Continued)

KEY TECHNOLOGIES	BRIEF DESCRIPTION	REFERENCES
Quick response code (QR code)	Refers to black and white square images that form a readable code that is usually used to store URLs to be read by devices such as tablets and smartphones.	(Okazaki et al., 2019)
Near field communication (NFC)	Refers to the contactless communication between electronic devices (smartphones, tablets, wearables, payment cards, etc.) to perform different tasks, including payments, sharing contacts, and passport IDs).	(Han et al., 2016)
Bluetooth (Bt)	Refers to wireless technology for exchanging data between electronic devices (mobile phones and computers) in a short-range network.	(Lin et al., 2022)
Global positioning system (GPS)	Refers to a radio-navigation system enabled by a set of satellites broadcasting navigation signals that give coordinates of an object in a given area.	(Khadhir et al., 2021)
Electronic data interchange (EDI)	Refers to the process of computer-to-computer exchange of business data in a standard format. This type of technology is used mainly on purchase activities between organizations (order transmitting, invoicing).	(Smith et al., 2020)
Radio-frequency identification (RFID)	Refers to radio wave technology that identifies, tracks and traces an object by using a tag.	(Ullah & Sarkar, 2020)

transaction costs through a peer-to-peer approach. Through simulation models, companies can save costs and time by testing different configurations to identify the best option before implementing a project. Using virtual reality and augmented reality, companies can improve and develop workers' skills through truly immersive education and training. Besides, artificial intelligence is considered one of the most popular technology of the digital transformation (Caiado et al., 2021; Cebollada et al., 2020; Collins et al., 2021; del Mar Roldán-García et al., 2017; De Oliveira et al., 2016; Dulebenets, 2019; Elshaer and Awad, 2020; Farhan et al., 2020; Hosseini and Ivanov, 2020; Kang et al., 2020; Kim et al., 2008; Klaib et al., 2021; Lee et al., 2018; Lima-Junior and Carpinetti, 2019; Paschen et al., 2019; Sabahi and Parast, 2020; Silva et al., 2021; Smiti and Soui, 2020; Sun and Medaglia, 2019; Syed et al., 2020).

2.4 SHORT CASE STUDY: WHY SHOULD COMPANIES INVEST IN KEY TECHNOLOGIES TO LEVERAGE THEIR DIGITAL TRANSFORMATION STRATEGY?

Smith is a senior manager in a large multinational company that explores and produces oil and gas. The company invests millions of dollars each year in research and development (R&D) and innovation. Despite its massive investments in information and communication technologies over the last three years, Smith has noticed that the company has not yet adopted key technologies like AI, BDA, or blockchain. In order to convince the company's board of directors to invest in these technologies, Smith hired a consulting firm to write a report on the benefits of these and other technologies specifically for an oil and gas company. You, as a consultant, must help Smith convince his company to adopt these and other key digital transformation technologies. In your plan, provide a strong argument about the pros and cons of the technologies and how they can support the company's R&D projects.

REFERENCES

Ajagekar, A., & You, F. (2019). Quantum computing for energy systems optimization: Challenges and opportunities. *Energy*, *179*, 76–89. https://doi.org/10.1016/j.energy.2019.04.186

Alzayat, A., & Lee, S. H. (Mark). (2021). Virtual products as an extension of my body: Exploring hedonic and utilitarian shopping value in a virtual reality retail environment. *Journal of Business Research*, *130*, 348–363. https://doi.org/10.1016/j.jbusres.2021.03.017

Beltagui, A., Kunz, N., & Gold, S. (2020). The role of 3D printing and open design on adoption of socially sustainable supply chain innovation. *International Journal of Production Economics*, *221*, 107462. https://doi.org/10.1016/j.ijpe.2019.07.035

Caiado, R. G. G., Scavarda, L. F., Gavião, L. O., Ivson, P., Nascimento, D. L. de M., & Garza-Reyes, J. A. (2021). A fuzzy rule-based industry 4.0 maturity model for operations and supply chain management. *International Journal of Production Economics*, *231*, 107883. https://doi.org/10.1016/j.ijpe.2020.107883

Cebollada, S., Payá, L., Flores, M., Peidró, A., & Reinoso, O. (2020). A state-of-the-art review on mobile robotics tasks using artificial intelligence and visual data. *Expert Systems with Applications*, *167,* 114195. https://doi.org/10.1016/j.eswa.2020.114195

Charro, A., & Schaefer, D. (2018). Cloud manufacturing as a new type of product-service system. *International Journal of Computer Integrated Manufacturing*, *31*(10), 1018–1033. https://doi.org/10.1080/0951192X.2018.1493228

Collins, C., Dennehy, D., Conboy, K., & Mikalef, P. (2021). Artificial intelligence in information systems research: A systematic literature review and research agenda. *International Journal of Information Management*, *60*, 102383. https://doi.org/10.1016/j.ijinfomgt.2021.102383

del Mar Roldán-García, M., García-Nieto, J., & Aldana-Montes, J. F. (2017). Enhancing semantic consistency in anti-fraud rule-based expert systems. *Expert Systems with Applications*, *90*, 332–343. https://doi.org/10.1016/j.eswa.2017.08.036

De Oliveira, E. M., Leme, D. S., Barbosa, B. H. G., Rodarte, M. P., & Alvarenga Pereira, R. G. F. (2016). A computer vision system for coffee beans classification based on computational intelligence techniques. *Journal of Food Engineering*, *171*, 22–27. https://doi.org/10.1016/j.jfoodeng.2015.10.009

Dulebenets, M. A. (2019). A delayed start parallel evolutionary algorithm for just-in-time truck scheduling at a cross-docking facility. *International Journal of Production Economics*, *212*, 236–258. https://doi.org/10.1016/j.ijpe.2019.02.017

Elshaer, R., & Awad, H. (2020). A taxonomic review of metaheuristic algorithms for solving the vehicle routing problem and its variants. *Computers and Industrial Engineering*, *140*, 106242. https://doi.org/10.1016/j.cie.2019.106242

Farhan, W., Talafha, B., Abuammar, A., Jaikat, R., Al-Ayyoub, M., Tarakji, A. B., & Toma, A. (2020). Unsupervised dialectal neural machine translation. *Information Processing and Management*, *57*(3), 102181. https://doi.org/10.1016/j.ipm.2019.102181

Fosso Wamba, S., Bawack, R. E., Guthrie, C., Queiroz, M. M., & Carillo, K. D. A. (2021). Are we preparing for a good AI society? A bibliometric review and research agenda. *Technological Forecasting and Social Change*, *164*, 120482. https://doi.org/10.1016/j.techfore.2020.120482

Fosso Wamba, S., Queiroz, M. M., Wu, L., & Sivarajah, U. (2020). Big data analytics-enabled sensing capability and organizational outcomes: Assessing the mediating effects of business analytics culture. *Annals of Operations Research*, 1–20. https://doi.org/10.1007/s10479-020-03812-4

Han, H., Park, A., Chung, N., & Lee, K. J. (2016). A near field communication adoption and its impact on Expo visitors' behavior. *International Journal of Information Management*, *36*(6), 1328–1339. https://doi.org/10.1016/j.ijinfomgt.2016.04.003

Hosseini, S., & Ivanov, D. (2020). Bayesian networks for supply chain risk, resilience and ripple effect analysis: A literature review. *Expert Systems with Applications*, *161*, 113649. https://doi.org/10.1016/j.eswa.2020.113649

Ivanov, D. (2020). Predicting the impacts of epidemic outbreaks on global supply chains: A simulation-based analysis on the coronavirus outbreak (COVID-19/SARS-CoV-2) case. *Transportation Research Part E: Logistics and Transportation Review*, *136*, 101922. https://doi.org/10.1016/j.tre.2020.101922

Ivanov, D. (2021). Digital supply chain management and technology to enhance resilience by building and using end-to-end visibility during the COVID-19 pandemic. *IEEE Transactions on Engineering Management*, 1–11. https://doi.org/10.1109/TEM.2021.3095193

Joerß, T., Hoffmann, S., Mai, R., & Akbar, P. (2021). Digitalization as solution to environmental problems? When users rely on augmented reality-recommendation agents. *Journal of Business Research*, *128*, 510–523. https://doi.org/10.1016/j.jbusres.2021.02.019

Kang, Z., Catal, C., & Tekinerdogan, B. (2020). Machine learning applications in production lines: A systematic literature review. *Computers and Industrial Engineering*, *149*, 106773. https://doi.org/10.1016/j.cie.2020.106773

Khadhir, A., Anil Kumar, B., & Vanajakshi, L. D. (2021). Analysis of global positioning system based bus travel time data and its use for advanced public transportation system applications. *Journal of Intelligent Transportation Systems*, *25*(1), 58–76. https://doi.org/10.1080/15472450.2020.1754818

Kim, M. C., Kim, C. O., Hong, S. R., & Kwon, I. H. (2008). Forward-backward analysis of RFID-enabled supply chain using fuzzy cognitive map and genetic algorithm. *Expert Systems with Applications*, *35*(3), 1166–1176. https://doi.org/10.1016/j.eswa.2007.08.015

Klaib, A. F., Alsrehin, N. O., Melhem, W. Y., Bashtawi, H. O., & Magableh, A. A. (2021). Eye tracking algorithms, techniques, tools, and applications with an emphasis on machine learning and Internet of Things technologies. *Expert Systems with Applications*, *166*, 114037. https://doi.org/10.1016/j.eswa.2020.114037

Lee, J., Jin, C., & Bagheri, B. (2017). Cyber physical systems for predictive production systems. *Production Engineering*, *11*(2), 155–165. https://doi.org/10.1007/s11740-017-0729-4

Lee, W. K., Leong, C. F., Lai, W. K., Leow, L. K., & Yap, T. H. (2018). ArchCam: Real time expert system for suspicious behaviour detection in ATM site. *Expert Systems with Applications*, *109*, 12–24. https://doi.org/10.1016/j.eswa.2018.05.014

Lima-Junior, F. R., & Carpinetti, L. C. R. (2019). Predicting supply chain performance based on SCOR ® metrics and multilayer perceptron neural networks. *International Journal of Production Economics*, *212*, 19–38. https://doi.org/10.1016/j.ijpe.2019.02.001

Lin, M. Y.-C., Nguyen, T. T., Cheng, E. Y.-L., Le, A. N. H., & Cheng, J. M. S. (2022). Proximity marketing and bluetooth beacon technology: A dynamic mechanism leading to relationship program receptiveness. *Journal of Business Research*, *141*, 151–162. https://doi.org/10.1016/j.jbusres.2021.12.030

Meindl, B., Ayala, N. F., Mendonça, J., & Frank, A. G. (2021). The four smarts of Industry 4.0: Evolution of ten years of research and future perspectives. *Technological Forecasting and Social Change*, *168*, 120784. https://doi.org/10.1016/j.techfore.2021.120784

Okazaki, S., Navarro, A., Mukherji, P., & Plangger, K. (2019). The curious versus the overwhelmed: Factors influencing QR codes scan intention. *Journal of Business Research*, *99*, 498–506. https://doi.org/10.1016/j.jbusres.2017.09.034

Paschen, J., Kietzmann, J., & Kietzmann, T. C. (2019). Artificial intelligence (AI) and its implications for market knowledge in B2B marketing. *Journal of Business and Industrial Marketing*, *34*(7), 1410–1419. https://doi.org/10.1108/JBIM-10-2018-0295

Queiroz, M. M., Fosso Wamba, S., Machado, M. C., & Telles, R. (2020). Smart production systems drivers for business process management improvement. *Business Process Management Journal*, *26*(5), 1075–1092. https://doi.org/10.1108/BPMJ-03-2019-0134

Queiroz, M. M., Pereira, S. C. F., Telles, R., & Machado, M. C. (2021). Industry 4.0 and digital supply chain capabilities. *Benchmarking: An International Journal*, 28(5), 1761–1782. https://doi.org/10.1108/BIJ-12-2018-0435

Sabahi, S., & Parast, M. M. (2020). The impact of entrepreneurship orientation on project performance: A machine learning approach. *International Journal of Production Economics*, 226, 107621. https://doi.org/10.1016/j.ijpe.2020.107621

Schwab, K. (2016). *The Fourth Industrial Revolution*. World Economic Forum.

Silva, A., Aloise, D., Coelho, L. C., & Rocha, C. (2021). Heuristics for the dynamic facility location problem with modular capacities. *European Journal of Operational Research*, 290(2), 435–452. https://doi.org/10.1016/j.ejor.2020.08.018

Smith, J. R., Yost, J., & Lopez, H. (2020). Electronic data interchange and enterprise resource planning technology in supply chain contracts. *Computers & Industrial Engineering*, 142, 106330. https://doi.org/10.1016/j.cie.2020.106330

Smiti, S., & Soui, M. (2020). Bankruptcy prediction using deep learning approach based on borderline SMOTE. *Information Systems Frontiers*, 22(5), 1067–1083. https://doi.org/10.1007/s10796-020-10031-6

Sun, T. Q., & Medaglia, R. (2019). Mapping the challenges of Artificial Intelligence in the public sector: Evidence from public healthcare. *Government Information Quarterly*, 36(2), 368–383. https://doi.org/10.1016/j.giq.2018.09.008

Syed, R., Suriadi, S., Adams, M., Bandara, W., Leemans, S. J. J., Ouyang, C., ter Hofstede, A. H. M., van de Weerd, I., Wynn, M. T., & Reijers, H. A. (2020). Robotic process automation: Contemporary themes and challenges. *Computers in Industry*, 115, 103162. https://doi.org/10.1016/j.compind.2019.103162

Tellez, E. S., Miranda-Jiménez, S., Graff, M., Moctezuma, D., Siordia, O. S., & Villaseñor, E. A. (2017). A case study of Spanish text transformations for twitter sentiment analysis. *Expert Systems with Applications*, 81, 457–471. https://doi.org/10.1016/j.eswa.2017.03.071

Ullah, M., & Sarkar, B. (2020). Recovery-channel selection in a hybrid manufacturing-remanufacturing production model with RFID and product quality. *International Journal of Production Economics*, 219, 360–374. https://doi.org/10.1016/j.ijpe.2019.07.017

Wamba, S. F., & Queiroz, M. M. (2022). Industry 4.0 and the supply chain digitalisation: A blockchain diffusion perspective. *Production Planning & Control*, 33(2–3), 193–210. https://doi.org/10.1080/09537287.2020.1810756

Wang, H., Ding, S., Wu, D., Zhang, Y., & Yang, S. (2019). Smart connected electronic gastroscope system for gastric cancer screening using multi-column convolutional neural networks. *International Journal of Production Research*, 57(21), 6795–6806. https://doi.org/10.1080/00207543.2018.1464232

Organizational Capabilities and Workers' Digital Skills

3.1 THE ROLE OF RESOURCES AND CAPABILITIES FOR DIGITAL TRANSFORMATION

Resources and capabilities in business management and related fields have been studied extensively (Barney, 2001). These two ingredients are the foundation of the well-known resource-based view theory (Barney, 1991, 2001), which posits that for firms to achieve a sustained competitive advantage, their resources should be valuable, rare, non-imitable, and non-substitutable. In this context, considering these resource characteristics, companies can leverage different capabilities.

In the digital transformation age, resources and capabilities continue to be a strong approach for organizations seeking to develop their strategies toward digitalization. Accordingly, digital transformation projects and strategies require different types of resources at different stages. The same is true regarding the capabilities that are activated/developed by the interaction of the resources.

While resources refer to the tangible or intangible assets/inputs that organizations have access to or control, a capability is a concept that refers to an organization's use of a set of resources to carry out its routine and strategic activities (Helfat & Peteraf, 2003). Companies traditionally organize their resources and capabilities into two categories. For instance, tangible resources can be physical like machinery, equipment, plants, distribution centers, and IT systems; intangible resources can include knowledge, leadership, top management support, and reputational capital. These types of resources can enable capabilities such as the ability to be responsive and flexible in meeting demand, agility, efficiency, alertness, predictive models, robustness, reconfiguration, and resilience.

Recently, the business analytics field positioned the resources and capabilities view from the three-part tangible, intangible, and human perspectives (Gupta & George, 2016; Kristoffersen et al., 2021). In this book, we extended this view to the digital transformation context. Figure 3.1 shows this approach. Accordingly, the digital transformation capability (DTC) of an organization has three types of resources: tangible, intangible, and human.

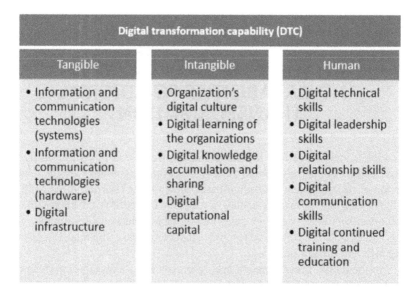

FIGURE 3.1   Digital transformation capability (DTC) framework. Adapted from Gupta and George (2016) and Kristoffersen et al. (2021).

Tangible resources are predominantly compounded by physical assets like IT infrastructure, which includes hardware, software, and equipment. Furthermore, tangible resources can include any type of resource owned or controlled by an organization that is related to its digital transformation strategy. For instance, if an organization's strategy is associated with an asset-light perspective, the company can use several cloud-service suppliers in order to minimize its investment in resources. Thus, the resources that the company is using are under its control. As such, they should appear on the tangible side of the DTC framework.

Intangible resources mainly refer to an organization's digital transformation strategy, its accumulated knowledge and the reputational capital relating to the use and achievements derived from the digital strategies. For example, the implementation of any strategy for digital transformation should consider the organization's digital culture. That is, the digital transformation should not be viewed as a single project but rather as a continuous approach to combining technologies, people, and processes, and the hierarchical levels should be conscious of this approach.

Human resources involve different skill sets. For example, these may be technical skills related to cutting-edge technologies (e.g., big data analytics, artificial intelligence, and simulation), continuous training to improve and acquire new digital skills, leadership/management, and relationship and communication skills. In the next two sections, we provide more details about the digital resources and capabilities of organizations and workers.

3.2 DIGITAL RESOURCES AND CAPABILITIES OF ORGANIZATIONS

In the organizational view, resources and capabilities (tangible, intangible, and human) interact with each other. In other words, the digital capability of an organization will depend on how efficiently and harmoniously this interaction occurs. There is no direct association regarding the number of resources organizations have; more resources do not necessarily make their digital transformation more successful. What matters is how effectively the organizations use available resources to create key digital capabilities and strategies for digital transformation.

In this way, an organization's digital resources play a fundamental role in enabling the key capabilities to support DTC. Digital resources refer to any type of information and communication technology (e.g., software, hardware, devices, and suppliers for digital services) that can be used by an organization and/or its partners through the network in order to not only support the conversion of analog tasks into a digitalized format but also raise the organization's digital knowledge share. Table 3.1 introduces this concept. For example, digital resources are divided into three pillars: ICTs, knowledge, and external digital resources. The digital transformation strategy of any organization should consider these pillars and employ efforts to improve them.

The first pillar refers to ICT digital resources. In this category, tangible resources like software, hardware, devices, and IT infrastructure predominate. The success of a company's digital transformation is not related to the number of these resources but instead depends on how they interact to add value to the organization's business. For instance, in a digital transformation project, software will add value only if it supports the improvement of an activity. That is, the software, hardware, device, etc. must ensure an incremental improvement.

The second category, knowledge of digital resources, refers to the organization's continuous efforts regarding training and learning on digital technologies and strategies to integrate them into the organization to create business value through the people. In this context, these efforts are essential for the development and improvement of knowledge sharing at all hierarchical levels of the organization. To target knowledge sharing, strategies concerning dissemination are crucial.

The third and last category is dedicated to external resource strategies. The first three types of resources are related to supplier services, co-creation, and research and development strategies. The last type of resource, similar to the internal dissemination of knowledge, refers to the external efforts of organizations to disseminate knowledge.

TABLE 3.1 Pillars of the digital resources of organizations

ICT DIGITAL RESOURCES	KNOWLEDGE DIGITAL RESOURCES	EXTERNAL DIGITAL RESOURCES
Software	Continuous learning	Suppliers for digital services
Hardware	Continuous training	Suppliers for digital co-creation
Devices	Knowledge sharing	Suppliers for R&D digital projects
IT infrastructure	Internal dissemination of the digital knowledge	External diffusion of the digital knowledge

3.3 DIGITAL RESOURCES AND CAPABILITIES OF WORKERS

With the current and unprecedented business model transformation, human activities are being rethought and remodeled (Hecklau et al., 2016; Queiroz et al., 2020). To survive in this complex and challenging scenario, new skills development and continuous learning are vital. Repetitive jobs will increasingly be done by machines, while humans will perform more activities related to analytics knowledge. Moreover, human–machine interaction will become more common in some jobs (Jarrahi, 2018), especially those related to operational activities.

Human–machine collaborations and interactions (Barreto et al., 2017) can have several benefits for organizations, including making tasks and activities safer and more efficient, reducing costs, and enhancing knowledge exchange. However, digital workers need a set of technical, methodological, social, and personal competencies (Hecklau et al., 2016; Ostmeier & Strobel, 2022).

Moreover, managers face challenges in planning and implementing strategies to support the organization's digital transformation. In this regard, they have to take into account the main stages of innovation dissemination: intention to adopt the new technology, adoption, and routinization (Martins et al., 2016), and the diffusion. Moreover, concerning other workers, activities related to cutting-edge technologies always need humans, but mainly to monitor, control, and apply data insights to the decision-making process.

One of the most important aspects concerning the human side in the view of digital resources is related to digital learning (Sousa & Rocha, 2019). It is clear that the digital skills of workers have the power to determine whether the digital transformation initiatives of an organization will be successful or not (Sousa & Rocha, 2019). As such, the constant improvement and development of new skills are critical for organizations wanting to maximize their achievements from digital transformation initiatives, such as innovation (Ciarli et al., 2021). Table 3.2 presents some of the most important workers' skills for digital transformation initiatives.

The skills on the communication side vary and include such strategic skills as management leadership and negotiation. Skills related to day-to-day activities like storytelling, presentation, and business writing are also in high demand for digital transformation initiatives. Additionally, emotional intelligence has been recognized as one of the most important skills for project success. Thus, communication skills and emotional intelligence are related to the nascent digital intelligence concept (Marnewick & Marnewick, 2021),

TABLE 3.2   Main workers' skills for digital transformation projects

COMMUNICATION	DATA LITERACY	EXTERNAL DIGITAL RESOURCES
Leadership	Analytics abilities	Strategic partnerships
Emotional intelligence	Problem-solving	Strategic cooperation
Storytelling	Critical thinking	Knowledge integration
Presentation	Cognitive abilities	Co-creation
Business writing	Innovative	Learning from partners
Negotiation	Creative	Collaboration with consultancy
Digital behavior	Programming abilities	Stakeholder monitoring to create opportunities

in which the technologies and digital skills that interact should not only support the organization's operations but should also have a positive impact on society as a whole, contributing to making progress and improving people's quality of life.

The data literacy dimension is shaped by technical and creative skills. From a technical point of view, skills related to analytics, problem-solving, and programming are some of the main skills for any digital transformation project. Skills related to cognitive abilities like innovation and creation are also essential to balance out the previous skills. Ultimately, external digital resources are also very important because, in many situations, organizations will not have all the necessary digital resources in-house. As a result, developing strategic partnerships and cooperation with their network is a critical point for the success of a digital transformation project. Doing so will imply knowledge integration, co-creation, and continuous learning.

3.4 SHORT CASE STUDY: HOW DO ORGANIZATIONS CREATE BUSINESS VALUE WITH RESOURCES AND CAPABILITIES IN THE DIGITAL TRANSFORMATION ERA?

A senior director of the company ZXWY, whose main product is a food delivery platform, is facing pressure from competitors and needs to improve the service level of its operation. Reports from the last year show that the main customer complaints are related to product quality (integrity), a lack of information concerning the order journey, delivery workers who do not understand the goods, and delivery delays. In a meeting of directors and managers, a supply chain director highlighted the urgency for a digital transformation project. However, most of the director's peers laughed and said that in a digitally native company, no digital transformation project was needed. Your task is to convince the board of directors and managers about the urgency of a digital transformation and how it can add value to the current operation and other projects at ZXWY.

REFERENCES

Barney, J. (1991). Firm resources and sustained competitive advantage. *Journal of Management, 17*(1), 99–120. https://doi.org/10.1177/014920639101700108

Barney, J. (2001). Resource-based theories of competitive advantage: A ten-year retrospective on the resource-based view. *Journal of Management, 27*(6), 643–650. https://doi.org/10.1016/S0149-2063(01)00115-5

Barreto, L., Amaral, A., & Pereira, T. (2017). Industry 4.0 implications in logistics: an overview. *Procedia Manufacturing, 13*, 1245–1252. https://doi.org/10.1016/j.promfg.2017.09.045

Ciarli, T., Kenney, M., Massini, S., & Piscitello, L. (2021). Digital technologies, innovation, and skills: Emerging trajectories and challenges. *Research Policy, 50*(7), 104289. https://doi.org/10.1016/j.respol.2021.104289

Gupta, M., & George, J. F. (2016). Toward the development of a big data analytics capability. *Information & Management, 53*(8), 1049–1064. https://doi.org/10.1016/j.im.2016.07.004

Hecklau, F., Galeitzke, M., Flachs, S., & Kohl, H. (2016). Holistic approach for human resource management in industry 4.0. *Procedia CIRP, 54*, 1–6. https://doi.org/10.1016/j.procir.2016.05.102

Helfat, C. E., & Peteraf, M. A. (2003). The dynamic resource-based view: capability lifecycles. *Strategic Management Journal, 24*(10), 997–1010. https://doi.org/10.1002/smj.332

Jarrahi, M. H. (2018). Artificial intelligence and the future of work: Human-AI symbiosis in organizational decision making. *Business Horizons, 61*(4), 577–586. https://doi.org/10.1016/j.bushor.2018.03.007

Kristoffersen, E., Mikalef, P., Blomsma, F., & Li, J. (2021). The effects of business analytics capability on circular economy implementation, resource orchestration capability, and firm performance. *International Journal of Production Economics, 239*, 108205. https://doi.org/10.1016/j.ijpe.2021.108205

Marnewick, C., & Marnewick, A. (2021). Digital intelligence: A must-have for project managers. *Project Leadership and Society, 2*, 100026. https://doi.org/10.1016/j.plas.2021.100026

Martins, R., Oliveira, T., & Thomas, M. A. (2016). An empirical analysis to assess the determinants of SaaS diffusion in firms. *Computers in Human Behavior, 62,* 19–33. https://doi.org/10.1016/j.chb.2016.03.049

Ostmeier, E., & Strobel, M. (2022). Building skills in the context of digital transformation: How industry digital maturity drives proactive skill development. *Journal of Business Research, 139,* 718–730. https://doi.org/10.1016/j.jbusres.2021.09.020

Queiroz, M. M., Fosso Wamba, S., Machado, M. C., & Telles, R. (2020). Smart production systems drivers for business process management improvement. *Business Process Management Journal, 26*(5), 1075–1092. https://doi.org/10.1108/BPMJ-03-2019-0134

Sousa, M. J., & Rocha, Á. (2019). Digital learning: Developing skills for digital transformation of organizations. *Future Generation Computer Systems, 91,* 327–334. https://doi.org/10.1016/j.future.2018.08.048

Barriers, Enablers, and the Digital Culture of the Organization

4

4.1 THE MOST COMMON BARRIERS TO DIGITAL TRANSFORMATION

Digital transformation initiatives, projects, and strategies come up against different types of behaviors that can reflect potential barriers. One of the main reasons for unsuccessful digital transformation projects can be due to **change management** (Melendez, 2021). Accordingly, any change in the current processes of established business activities over the years can impact the active resistance of workers at different hierarchical levels. These dynamics require substantial efforts from all decision levels to communicate the benefits of digital transformation initiatives.

In this regard, **miscommunication** between team members (Bhattacharya & Momaya, 2021) and about the benefits and goals of the digital transformation

DOI: 10.1201/9781003226468-4

can have negative impacts, such as implementation delays or, in many cases, complete project failure. Determining an effective channel of communication in the company and with stakeholders' involvement about where the company is and where it wants to go with its digital transformation project are fundamental to overcoming this type of barrier.

Another significant barrier is related to **resources** (Fischer et al., 2020). As highlighted in the previous chapter, resources can be tangible, intangible, and human (Kristoffersen et al., 2021). When viewing resources in this way, companies fail mainly when they do not understand the need for continuous development of resources for their digital transformation initiative—for instance, when it comes to human resources, continuous training, and partnerships with other companies are essential. With regard to tangible and intangible resources, companies should develop a mix of in-house and external resources with key suppliers.

In addition, **nonrealistic costs and benefits** also can be huge barriers (Correani et al., 2020). Some companies are unsuccessful in their digital transformation because they do not develop a realistic cost projection, with managers tending to have overly optimistic expectations about the required time to fulfill a digital transformation project. Consequently, the cost flow is often not well defined. This is a major error that happens because a digital transformation is not a traditional project with a specific start and end date. It requires companies to ensure a continuous flow of investment to keep up with advances and integrate into them their business. Similarly, organizations often take an overly rosy view of the potential benefits. Companies will not automatically achieve benefits simply by adopting digital technologies or revising their business model.

Another important barrier that is common in digital transformation projects is the **legacy systems** of the organizations and their network (Saarikko et al., 2020). The number of legacy systems (outdated systems) can lead to difficulties and delays in integrating them with the new solutions. Thus, companies that go ahead with their digital transformation initiatives when they have a substantial number of legacy systems will not be able to achieve a full digital transformation process; this results in the business model's operations and efficiency being negatively impacted.

Top management support is one of the most important aspects of digital transformation initiatives (Weber et al., 2022; Wrede et al., 2020). The management is responsible for leading change by integrating the both internal and external resources (tangible, intangible, and human) in a harmonious way. When this does not happen, barriers due to a **lack of top management support** emerge. As a result, this type of barrier can jeopardize the entire project. Good leadership is essential to guide the organization and its partners on the required resources, priorities, costs, and investments, among other activities.

A **lack of workers' skills** is a typical barrier that most organizations face in their digital transformation projects. Workers' skills are already known to play a fundamental role in the success of digital transformation projects (Blanka et al., 2022). Such skills may be technical in nature (e.g., analytics, AI, and programming) or non-technical/soft skills (e.g., emotional intelligence, communication, and creativity). Companies should also be aware that talented workers need an environment of continuous innovation and motivation, supported by continuous training, to address digital transformation challenges and implement changes.

Employee commitment to digital transformation projects exerts an important influence on the success of such projects (Cichosz et al., 2020). Thus, a **lack of commitment** is also a significant barrier. Commitment to digital transformation initiatives not only encompasses all hierarchical levels within the organization, but also surpasses company boundaries by integrating key partners. Because of this, a low level of commitment can negatively impact the implementation of the action, resulting in increased project costs and delays.

Collaboration between partners can unlock valuable and decisive insights for digital transformation initiatives (Busulwa et al., 2022). A lack of collaboration between the organization's departments—for instance, misalignment between IT and business departments—can negatively impact the operations, creating delays, increasing costs, and causing failures in implementation. Consequently, the digital transformation initiative cannot generate the expected business value.

Finally, for digital transformation initiatives to be successful, a company must have a clear vision (Tijan et al., 2021). This implies that the company has a clear strategy and targets and will consider any changes in the main stakeholders. With a clear vision, organizations can plan where they should go and how to get there in a more effective way. A **lack of vision** can emerge as a major barrier to a digital transformation project. The main issue is miscommunication about the organization's targets and objectives and which resources will be mobilized. Generally speaking, all hierarchical levels either do not know or have serious difficulties understanding where the company is going.

Table 4.1 summarizes the main barriers to digital transformation projects and offers some strategies to overcome them.

4.2 ENABLERS OF DIGITAL TRANSFORMATION INITIATIVES

In digital transformation projects, enablers are recognized as the drivers of digital transformation initiatives. They are related to a set of capabilities in the form of applications or services (Schallmo et al., 2017) that are used to

TABLE 4.1   Main barriers in digital transformation projects and strategies to overcome them

BARRIERS	STRATEGIES TO OVERCOME THE BARRIERS
Change management	Development of communication plans at all hierarchical levels about the benefits derived from the digital transformation.
Miscommunication	Development of a powerful communication channel that accounts for the main stakeholders to share the company's plan about where it wants to go with its project and the efforts required by the partners.
Lack of resources	Continuous development and improvement of resources, including in-house and with key partners.
Nonrealistic costs and benefits	Development of detailed cash flow for different time frames. Utilize benchmarks from digital transformation initiatives to help identify the most common benefits at each stage of the project.
Legacy systems	Development of continuous actions to modernize the systems. This involves a careful plan to replace the legacy systems with modern solutions. It should be a smooth process.
Lack of top management support	Identification of the skilled leaders (internal and or external) and continuous training and motivation for driving the changes. Implementation of reward policies based on the achievements of the digital initiatives.
Lack of workers' skills	Implementation of a continuous training policy to develop and improve technical skills and soft skills.
Lack of commitment	Development of an efficient channel to communicate why the change is essential to the company. Implementation of a reward policy at all employee levels based on their performance.
Lack of collaboration	Development of a continuous process to check the alignment of the objectives between the teams. More in-depth integration between IT and business teams.
Lack of vision	Development of a clear vision from the top that is shared with all company departments, as well as with the stakeholders. It is essential to share the status of the project, benefits, the changes that are required and expected, how the partners will be affected, how to measure the success, etc.

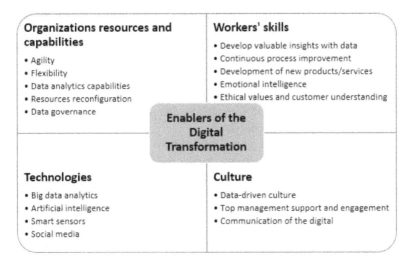

FIGURE 4.1   Main enablers of the digital transformation.

transform a company's business model and drive digital projects forward. The most common enablers are an organization's resources and capabilities, workers' skills, technologies, and culture (Bain & Company, 2018). Figure 4.1 shows a framework most popular enablers and how they interact.

The enablers in Figure 4.1 emphasize the importance of the interaction between different types of resources that are essential to the digital transformation projects and initiatives. Accordingly, four types of enablers should be considered by the organization: resources and capabilities, workers' skills, technologies, and culture.

4.3 CULTURE AS A SECRET INGREDIENT IN DIGITAL TRANSFORMATION

Organizational culture refers to an organization's values and beliefs, which are used to support their routines and operations (Liu et al., 2010). In the age of the digital transformation phenomenon, a variation of organizational culture—digital culture—has emerged and provides an explanation for the digital transformation phenomenon. In this regard, and considering Figure 4.1, digital culture refers to the organization's behavior, values, engagement with data-driven practices, leadership support and communication of the initiatives, and goals of digital transformation practices.

Digital culture is basically shaped by the following aspects: collaboration, data-driven practices, customer-centric focus, innovation, and purpose (DuBridge & Tippens, 2021; World Economic Forum, 2021). From this perspective, the collaborative dimension reflects the need for strategic collaborations in the organization's network in order to create strong partnerships to develop innovative solutions for its customers. The data-driven practices must support the organization with insights and predictive analysis to aid the complex decision-making process.

In the customer-centric view, organizations need to make an effort to maximize the customer relationship in order to develop customized solutions. In terms of innovation, organizations must operate in an atmosphere that supports continuous new ideas and projects from any hierarchical level. Finally, the purpose aspect emphasizes that a company's digital culture should consider society as a whole by acting according to ethical and sustainable practices, as well as supporting diversity, equity, and inclusion.

4.4 SHORT CASE STUDY: HOW CAN CULTURE OVERCOME BARRIERS IN DIGITAL TRANSFORMATION STRATEGIES?

A multinational computer company that has business units across the global recently launched a digital transformation project for all of its subsidiaries (which have operations on all continents). Two years after the project kickoff, half of the initiatives were interrupted. A quarter of the initiatives successfully achieved the expected outcomes (improved business process activities, more value-added for customers, reduced operational costs, etc.); another quarter is still active but suffering from severe delays and rising costs. After visiting the unsuccessful subsidiaries, an expert from a consulting firm identified organizational culture as one of the major issues. More specifically, the consultant found that these subsidiaries have inadequate levels of digital culture. It is interesting to note that the company had both successful and unsuccessful implementation projects on all continents. Furthermore, the top managers of subsidiaries that achieved the planned outcomes with the digital transformation project told the other subsidiaries that there was no need to continue investing in digital transformation actions since they had already achieved their objectives. As a result, the company is planning to stop its investments in resources and capabilities because they are too expensive. In what ways did the company fail in implementing its global digital transformation project? What are the possible consequences of its decision to stop its investments? How can you convince the company to change its decision?

REFERENCES

Bain & Company. (2018). *Digital Transformation Roadmap: Enablers*. https://www.bain.com/insights/digital-transformation-roadmap-enablers/

Bhattacharya, S., & Momaya, K. S. (2021). Actionable strategy framework for digital transformation in AECO industry. *Engineering, Construction and Architectural Management*, 28(5), 1397–1422. https://doi.org/10.1108/ECAM-07-2020-0587

Blanka, C., Krumay, B., & Rueckel, D. (2022). The interplay of digital transformation and employee competency: A design science approach. *Technological Forecasting and Social Change*, 178, 121575. https://doi.org/10.1016/j.techfore.2022.121575

Busulwa, R., Pickering, M., & Mao, I. (2022). Digital transformation and hospitality management competencies: Toward an integrative framework. *International Journal of Hospitality Management*, 102, 103132. https://doi.org/10.1016/j.ijhm.2021.103132

Cichosz, M., Wallenburg, C. M., & Knemeyer, A. M. (2020). Digital transformation at logistics service providers: Barriers, success factors and leading practices. *The International Journal of Logistics Management*, 31(2), 209–238. https://doi.org/10.1108/IJLM-08-2019-0229

Correani, A., de Massis, A., Frattini, F., Petruzzelli, A. M., & Natalicchio, A. (2020). Implementing a digital strategy: Learning from the experience of three digital transformation projects. *California Management Review*, 62(4), 37–56. https://doi.org/10.1177/0008125620934864

DuBridge, S., & Tippens, B. (2021). *Top-Down or Bottom-Up? How to Cultivate a Digital Culture in Your Organization*. World Economic Forum. https://www.weforum.org/agenda/2021/10/innovation-as-a-test-case-for-digital-collaboration/

Fischer, M., Imgrund, F., Janiesch, C., & Winkelmann, A. (2020). Strategy archetypes for digital transformation: Defining meta objectives using business process management. *Information & Management*, 57(5), 103262. https://doi.org/10.1016/j.im.2019.103262

Kristoffersen, E., Mikalef, P., Blomsma, F., & Li, J. (2021). The effects of business analytics capability on circular economy implementation, resource orchestration capability, and firm performance. *International Journal of Production Economics*, 239, 108205. https://doi.org/10.1016/j.ijpe.2021.108205

Liu, H., Ke, W., Wei, K. K., Gu, J., & Chen, H. (2010). The role of institutional pressures and organizational culture in the firm's intention to adopt internet-enabled supply chain management systems. *Journal of Operations Management*, 28(5), 372–384. https://doi.org/10.1016/j.jom.2009.11.010

Melendez, C. (2021). *Five Barriers to Digital Transformation and How to Overcome Them*. Forbes Technology Council. https://www.forbes.com/sites/forbestech-council/2021/04/01/five-barriers-to-digital-transformation-and-how-to-overcome-them/?sh=782112bcb112

Saarikko, T., Westergren, U. H., & Blomquist, T. (2020). Digital transformation: Five recommendations for the digitally conscious firm. *Business Horizons*, *63*(6), 825–839. https://doi.org/10.1016/j.bushor.2020.07.005

Schallmo, D., Williams, C. A., & Boardman, L. (2017). Digital transformation of business models—best practice, enablers, and roadmap. *International Journal of Innovation Management*, *21*(08), 1740014. https://doi.org/10.1142/S136391961740014X

Tijan, E., Jović, M., Aksentijević, S., & Pucihar, A. (2021). Digital transformation in the maritime transport sector. *Technological Forecasting and Social Change*, *170*, 120879. https://doi.org/10.1016/j.techfore.2021.120879

Weber, E., Büttgen, M., & Bartsch, S. (2022). How to take employees on the digital transformation journey: An experimental study on complementary leadership behaviors in managing organizational change. *Journal of Business Research*, *143*, 225–238. https://doi.org/10.1016/j.jbusres.2022.01.036

World Economic Forum. (2021). *Digital Culture: The Driving Force of Digital Transformation*. https://www3.weforum.org/docs/WEF_Digital_Culture_Guidebook_2021.pdf

Wrede, M., Velamuri, V. K., & Dauth, T. (2020). Top managers in the digital age: Exploring the role and practices of top managers in firms' digital transformation. *Managerial and Decision Economics*, *41*(8), 1549–1567. https://doi.org/10.1002/mde.3202

The Role of the Digital Supply Chain

5

5.1 BASIC ASPECTS OF THE SUPPLY CHAIN

Supply chain management (SCM) is a multidisciplinary field that is recognized as one of the last frontiers for organizations seeking a competitive advantage in a complex contemporary world. Although SCM has a long tradition in business management and related fields, there is no unified concept (Mentzer et al., 2001). This book adheres to the definition provided by the (Council of Supply Chain Management Professionals [CSCMP], 2022):

> Supply chain management encompasses the planning and management of all activities involved in sourcing and procurement, conversion, and all logistics management activities. Importantly, it also includes coordination and collaboration with channel partners, which can be suppliers, intermediaries, third party service providers, and customers. In essence, supply

DOI: 10.1201/9781003226468-5

chain management integrates supply and demand management within and across companies.

Based on this definition, SCM includes strategic, tactical, and operational activities. Accordingly, SCM involves developing strategic relationships with key members of an organization's network (Gibson et al., 2005) as well as optimizing established collaborations. One of the most important aspects of SCM deals with the integration between companies and how they interact in order to meet customer demand. From this perspective, digital transformation has created new possibilities for operations and SCM, as well as new challenges (Holmström et al., 2019; Seyedghorban et al., 2020). In the next section, the integration of the supply chain with digital transformation will be explored in greater depth.

5.2 FUNDAMENTALS OF THE DIGITAL SUPPLY CHAIN

With the advances and spread of digital transformation concepts, SCM is also shifting toward a more digitalized operations model. In this landscape, the digital supply chain (DSC) can be defined as

> an intelligent best-fit technological system that is based on the capability of massive data disposal and excellent cooperation and communication for digital hardware, software, and networks to support and synchronize interaction between organizations by making services more valuable, accessible and affordable with consistent, agile and effective outcomes.
>
> *(Büyüközkan & Göçer, 2018, p. 165)*

Following this definition and in connection with the TOP framework, this book adopts the following definition of a DSC: DSC refers to the integration of cutting-edge technologies, people, and organizations, by using the latest know-how and best practices, in order to comprehensively support the relationships and cooperation between organizations and their supply chains, with the objective of creating value for stakeholders (i.e., customers, organizations, government, and society) and contributing to the social good in accordance with ethical values.

In this definition, the central point is the creation of value by contemplating society's ethical values (i.e., fairness, responsibility, honesty, and integrity). For instance, based on a DSC perspective, if a company wants to satisfy customers by offering a distinct service level, the organization and its

Integration	Relationship and cooperation	Value creation	Ethical values
•Technology •Organizations •People •Know-how	•Organizations •Cooperation of the internal departments •Supply Chains •Cooperation of the external partners	•Customers •Organizations •Government •Society	•Diversity, equity, and inclusion (DEI) •Fairness •Responsibility •Honesty •Integrity

FIGURE 5.1   Dimensions of the digital supply chain.

supply chains should minimize carbon emissions not only from transportation activities but in all stages of SCM by adopting practices and policies of diversity, equity, and inclusion (DEI) throughout their processes, supported by technology, people and the organization's values. Figure 5.1 emphasizes the main dimensions of the DSC definition. Thus, four dimensions are highlighted: integration (TOP framework), relationship and cooperation (organizations and supply chains), value creation (for stakeholders), and ethical values (e.g., DEI).

5.2.1 Digital Supply Chain: The Integration Dimension

This dimension highlights the importance of the TOP framework as one of the pillars to enable the development of DSCs. The technology approach is important, but, as highlighted in previous chapters, simply using cutting-edge technologies without tying them into a larger digital transformation strategy does not constitute digital transformation. In operational and supply chain approaches, technologies play an essential role in supporting the activities (operational levels), as well as shoring up decisions from the strategic planning point of view. In this regard, technology management requires substantial attention from the different supply chain teams, who must consider the adequate type of technology for each activity and how those technologies contribute or not to creating/increasing value for the organization and customers.

From this perspective, the other categories from the TOP framework—organizations and people interacting with key technologies—should provide adequate support for the supply chain shift for a digital model operation. This means that organizations need to provide adequate resources (i.e., financial, culture, infrastructure, and top management support) and highly skilled people in operational and supply chain topics to support a smooth transformation.

Take, for example, a company has an inefficient process related to follow-up activities, which costs a significant amount of money and minimizes the value to customers because of delays in information. To digitalize these activities, the company should consider using blockchain, IoT, RFID, and sensors in an integrated way, but adopting the technology is only one of the first steps. The company needs constant support from the organization (decision-makers), workers who receive continuous training to improve the process, and deep integration with the key member(s) of the supply chain. To be sustainable, this loop requires generating internal and external knowledge sharing about value creation and improvement.

5.2.2 Digital Supply Chain: The Relationship and Cooperation Dimension

In the relationship and cooperation dimension, to enable the development of a DSC, two main approaches are required—cooperation among the organization's internal departments and cooperation with the external supply chain partners. The organization's internal departments (e.g., finance, production, purchasing, human resources, logistics, operations, and supply chains) need to operate in a harmonious way, taking into account the needs and impacts (positive/negative) that the digital transformation will generate. To enable the digital transformation in the supply chains, the organization's different departments must be highly integrated.

Accordingly, this dimension shows the need for an in-depth view of the digital strategy of the organization, considering all hierarchical levels. As such, this dimension implies the development of high levels of awareness about the current position of the organization and where it expects the organization to go in the coming years, supported by digital transformation. To develop a DSC, the aforementioned departments need to improve the communication process between the teams, promoting adequate support to acquire and develop new capabilities through the available resources. In addition, the knowledge acquired by one department should be quickly shared with the other departments.

With regard to the relationship and cooperation with external partners, this dimension is predominantly related to the interaction with supply chain members. Similar to the dynamics between internal departments, organizations need to ensure a well-connected channel of communication to engage the key partners about the business and processes and how they will be impacted in the short, middle, and long terms. Furthermore, there should be a clear understanding of what is expected from the partners and what will be gained. The resources and capabilities of partners should be integrated with the organization in order to leverage the DSCs. Here again, knowledge produced by any partner should be shared throughout the network.

5.2.3 Digital Supply Chain: The Value Creation Dimension

The value creation dimension considers customers, organizations, government, and society. The DSC can integrate all these stakeholders into its business model and operations at the operational, tactical, and strategic levels. For instance, from an operational standpoint, the DSC can make communication with customers easier and smoother, providing real-time information and gathering real-time feedback.

Regarding the tactical standpoint, managers can, for example, use artificial intelligence and big data analytics to analyze a large data set from different sources from these stakeholders in order to support the creation of new products and services and improve the user experience. From a strategic standpoint, the use of cutting-edge technologies, supported by a focus on digital culture with highly skilled workers, can establish strategic partnerships with other organizations in the supply chain and with government stakeholders, thus adding more value to all involved. For example, by using digital twins (Ivanov & Dolgui, 2021), organizations can get real-time information about these stakeholders. This is essential to support strategic decisions, which will impact society across all time frames.

5.2.4 Digital Supply Chain: The Ethical Values Dimension

The ethical values dimension of the DSC needs to be supported by the DEI view (Johnson & Chichirau, 2020). From this standpoint, for organizations to achieve a fully DSC, they need to consider the differences (i.e., ethnicity, religion, gender, underrepresented groups, etc.) not only in their company but throughout their networks. To promote this view, the digital culture of the organization can support equity and inclusion actions by leveraging social justice, fairness, responsibility, honesty, and integrity for all groups that the organization may impact. In this way, the DSC can create additional value for workers, customers, and society through more transparent practices, bolstered by cutting-edge technologies, a strong digital culture, and top management support.

5.3 A FRAMEWORK FOR THE DIGITAL SUPPLY CHAIN

In order to provide a more integrated view of the previous foundations of the DSCs, Figure 5.2 shows a basic framework. The dimensions (integration, relationship and cooperation, value creation, and ethical values) are the central point of the framework. Their interaction depends on the creation and success of the DSC. From the strategic perspective, these dimensions are reinforced by the digital culture of the supply chain organizations, top management support, and leadership. From the tactical and operational standpoint, these relationships, when combined with the use of cutting-edge technologies, should be capable of creating products and services with high added value, providing an in-depth experience and excellent service level to customers. This high added value should be reflected through a good society, inequality minimization, carbon neutrality, environmental sustainability, and disruption minimization provoked by different types of crises (i.e., pandemics, war risks, climate changes, financial, humanitarian, etc.).

FIGURE 5.2   Framework of the digital supply chain.

5.4 SHORT CASE STUDY: THE DIGITAL SUPPLY CHAIN AND THE LAST FRONTIER OF ADDED VALUE

Consider the following situation: Company ABCD is a global fashion industry leader. More specifically, ABCD is a global fashion brand that is analyz-
ing whether opening a new business unit specialized in fast fashion would
be a good idea. The fast fashion operation will require greater efficiency in
mass-production strategies, via a low-cost operation and high dependence
on logistics and supply chains. According to a consulting firm, ABCD will
need a DSC structure to implement the fast fashion project. Although the
digital transformation project began three years ago, the company is not
confident in the project having a positive outcome. More specifically, the
digital transformation project is yet to be integrated into the supply chain.
Furthermore, the company's management board has reported that the digital
transformation project has nearly spent its entire budget. In this challenging
scenario, the supply chain director has been asked to launch a DSC project so
the company can enter the fast fashion market as soon as possible. Based on
the "Framework of the digital supply chain," help the director convince the
stakeholders about the benefits of implementing a DSC project to support the
fast fashion project.

REFERENCES

Büyüközkan, G., & Göçer, F. (2018). Digital supply chain: Literature review and a
    proposed framework for future research. *Computers in Industry*, *97*, 157–177.
    https://doi.org/10.1016/j.compind.2018.02.010
Council of Supply Chain Management Professionals (CSCMP). (2022). *CSCMP
    Supply Chain Management Definitions and Glossary*. https://cscmp.org/
    CSCMP/Educate/SCM_Definitions_and_Glossary_of_Terms.aspx
Gibson, B. J., Mentzer, J. T., & Cook, R. L. (2005). Supply chain management: The
    pursuit of a consensus definition. *Journal of Business Logistics*, *26*(2), 17–25.
    https://doi.org/10.1002/j.2158-1592.2005.tb00203.x
Holmström, J., Holweg, M., Lawson, B., Pil, F. K., & Wagner, S. M. (2019). The digi-
    talization of operations and supply chain management: Theoretical and meth-
    odological implications. *Journal of Operations Management*, *65*(8), 728–734.
    https://doi.org/10.1002/joom.1073

Ivanov, D., & Dolgui, A. (2021). A digital supply chain twin for managing the disruption risks and resilience in the era of Industry 4.0. *Production Planning & Control*, *32*(9), 775–788. https://doi.org/10.1080/09537287.2020.1768450

Johnson, M. P., & Chichirau, G. R. (2020). Diversity, equity, and inclusion in operations research and analytics: A research agenda for scholarship, practice, and service. *INFORMS TutORials in Operations Research*, 1–38. https://doi.org/10.1287/educ.2020.0214

Mentzer, J. T., DeWitt, W., Keebler, J. S., Min, S., Nix, N. W., Smith, C. D., & Zacharia, Z. G. (2001). Defining supply chain management. *Journal of Business Logistics*, *22*(2), 1–25. https://doi.org/10.1002/j.2158-1592.2001.tb00001.x

Seyedghorban, Z., Tahernejad, H., Meriton, R., & Graham, G. (2020). Supply chain digitalization: Past, present and future. *Production Planning & Control*, *31*(2–3), 96–114. https://doi.org/10.1080/09537287.2019.1631461

Digital Supply Chain Capabilities

6

6.1 DIGITAL SUPPLY CHAIN RESOURCES

In the digital supply chain, resources play an essential role in shaping key capabilities. On the one hand, the strategic field defines resources as all types of assets (tangible and intangible), including capabilities, that an organization manages, with the objective of supporting organizational efficiency and efficacy (Barney, 1991; Wernerfelt, 1984). On the other hand, we provide the following definition for resources in the digital supply chain context. Resources include technology, the organization's infrastructure, values and beliefs, and people, all of which operate in a cooperative and integrated manner, sharing ethical values to support internal and external value creation.

Capabilities in this context are considered a result of the interaction of resources. Accordingly, with regard to the digital supply chain, capabilities are defined as the organization's abilities to employ and integrate their available resources (internal and external) through the supply chains, with the objective of supporting business value creation. Thus, the quality of the capabilities of the digital supply chain is determined by resource interaction and management. The following subsections provide more details about the resources and capabilities of the digital transformation.

6.1.1 Basic Resources of the Digital Supply Chain

Following on from the definition of digital supply chain resources, from a technology perspective, the basic resources are as follows:

- Cutting-edge technologies (Meindl et al., 2021): artificial intelligence, big data analytics, IoT, simulation, etc.;
- Internal and external infrastructure to support the technologies: hardware, cloud partners, etc.

Regarding the organization's infrastructure, values and beliefs, the following resources are the most common:

- Tangible assets: machinery, equipment, facilities, buildings, distribution centers, vehicles, etc.;
- Intangible assets: digital culture, data-driven culture, top management support, leadership, commitment to stakeholders, sustainability policies, etc.

Finally, with regard to people, the main resources of the digital supply chain are as follows:

- Communication;
- Emotional intelligence;
- Data literacy;
- Problem-solving and decision-making;
- Teamwork and collaboration;
- Innovation and creativity.

6.2 CAPABILITIES OF THE DIGITAL SUPPLY CHAIN

As highlighted previously, the interaction between the resources of the digital supply chain enables the emergence of the capabilities. The main digital supply chain capabilities are workers' digital skills, smart production, digital warehousing, smart transportation, supplier integration, and customer collaboration. These capabilities are essential to the operations, and today's supply chains face the demands and challenges imposed by the environment.

6.2.1 Workers' Digital Skills Capabilities

Worker's digital skills capabilities refer to employees' abilities and behaviors with regard to using cutting-edge technologies to add value to the business of the organization and supply chain. Nowadays, workers' skills are essential for any type of organization. Taking into account supply chain dynamics and complexities, digital skills development plays a critical role in digital supply chains. Accordingly, the following capabilities can be highlighted as the most representative. It is important to note that due to the dynamics and rapid advances of technologies, some digital skills also change.

- *Data analytics capabilities* refer to skills that are used to identify different data sources and to integrate, store, and analyze that data using data science approaches (programming packages, statistical and mathematical methods). It also includes robust storytelling in presentations to stakeholders.
- *Process remodeling capabilities* refer to skills of operations and supply chain that are combined with business and management views, supported by cutting-edge technologies, which are used to remodel the analog process and convert it to digital, with the objective of adding more value to the organization and its supply chain.
- *Digital emotional intelligence and behavior* refer to skills related to relationship behaviors in the organization and supply chains. It involves ethical values in using digital technologies to support the business activities, as well as aspects related to the behavior of workers on social media.

6.2.2 Smart Production Capabilities

Smart production capabilities refer to the use of technologies that enable production systems to adopt a smart paradigm (Queiroz et al., 2020). This capability includes the use of additive manufacturing, cyber-physical systems, sensors, cloud chain, artificial intelligence, simulation, big data analytics, etc. in the production systems, with the objective of supporting product design in a more responsive and flexible manner to meet the need for mass customization. Furthermore, aspects related to maintenance, quality, and scheduling are significantly improved when this capability is developed.

6.2.3 Digital Warehousing Capabilities

Digital warehousing capabilities refer to the automated warehouses supported by cutting-edge technologies, which execute a set of processes and activities with minimum human interventions. In recent years, technologies such as AR/VR have become popular in operations and supply chains (Akbari et al., 2022). For instance, AR/VR can be used in order picking operations (Winkelhaus et al., 2021).

Other technologies also contribute to leveraging digital warehousing capabilities. Examples include collaborative robots (cobots), automated guided vehicles, AI, IoT, and mobile robot fulfillment systems, among others (Winkelhaus et al., 2021). The greater the demand on the warehouse processes and activities, the greater the warehouse's need to become more digital. In addition, with digital twin approaches, organizations/supply chains can create a digital copy of the warehouse (Leung et al., 2022) to then analyze the dynamics of the warehouse operations in real time. It is important to note that to be a digital warehouse, an organization and its supply chains do not acquire all types of technologies to automate their processes. The capabilities should be compounded by a set of key technologies that can add more value based on the demand, flow, supply chain structure, costs, etc.

6.2.4 Smart Transportation Capabilities

Smart transportation capabilities refer to the use of cutting-edge and established technologies to support traditional activities of goods transport (i.e., traceability, vehicle routing, freight management, fleet management, crew composition, etc.). For instance, with blockchain technologies (Queiroz et al., 2021), organizations and their supply chains can improve their traceability capabilities, monitoring the entire journey of the goods in the network. In a digital supply chain, the smarter transportation activities are, the more transparent and reliable the supply chain will be.

Furthermore, with the use of smart city infrastructure, transportation could be smarter and more autonomous by using the information and communication technologies offered by cities (Yan et al., 2020). For example, GPS, RFID, sensors, cloud chain, etc. can enable capabilities related to congestion monitoring, dynamic routes, and communication with other vehicles.

6.2.5 Supplier Integration Capabilities

Supplier integration capabilities refer to the integration enabled by cutting-edge technologies of organizations with their suppliers throughout the network. In this regard, technologies such as blockchain (Wamba & Queiroz, 2020) can leverage a greater degree of supplier integration across the supply chains, thus supporting enhanced information sharing and improving transaction transparency. Moreover, by using cutting-edge technologies in activities involving suppliers, a purchasing 4.0/procurement 4.0 platform can be implemented (Bienhaus & Haddud, 2018; Gottge et al., 2020) to integrate and manage different types of processes and benefit the supply chain members with more agility in transactions.

6.2.6 Customer Collaboration Capabilities

Customer collaboration capabilities refer to the integration of customers by providing offline/online feedback about their experience with a product/service of the organization. The integration is also supported by the application of different technologies. For example, customers can provide feedback using social media about their experience with product delivery, while AI technologies let customers with smartphones track their goods in real time. Companies can also use blockchain technologies to improve the transparency and provenance of the products (Montecchi et al., 2019).

In addition, other approaches like wearables/AR can provide supply chains with real-time information about the customer experience with a product; this allows customers to be co-creators in the supply chains (Gupta et al., 2018). In this co-creation view, approaches using software-as-a-service (SaaS) to enable real-time information from customers (Mujahid Ghouri et al., 2021) can be a valuable resource for operations and supply chain activities and strategies.

6.3 SUMMARIZING THE DIGITAL SUPPLY CHAIN CAPABILITY

In summary, Figure 6.1 provides an integrated view of the main digital supply chain capabilities (workers' digital skills, smart production, digital warehousing, smart transportation, supplier integration, and customer collaboration).

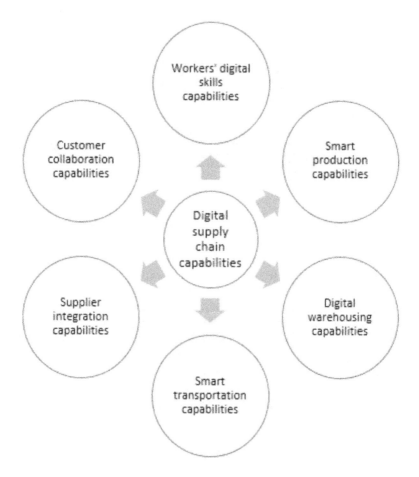

FIGURE 6.1   Integrated view of the digital supply chain capabilities.

6.4 SHORT CASE STUDY: DIGITAL SUPPLY CHAIN CAPABILITY

A supply chain director from a global grocery store chain is facing several problems. The main challenges are related to an increased number of items being sold, which leads to difficulties in receiving, storage, and picking/packing operations. Additional issues have emerged in recent months, such as stockouts and expired products. Currently, for these activities, the organization uses only warehouse management software. Based on the digital supply chain capabilities approach, how can the main capabilities (workers' digital skills, smart production, digital warehousing, smart transportation, supplier integration, and customer collaboration) be used to support the director to minimize these challenges?

REFERENCES

Akbari, M., Ha, N., & Kok, S. (2022). A systematic review of AR/VR in operations and supply chain management: Maturity, current trends and future directions. *Journal of Global Operations and Strategic Sourcing*, 1–32. https://doi.org/10.1108/JGOSS-09-2021-0078

Barney, J. (1991). Firm resources and sustained competitive advantage. *Journal of Management*, *17*(1), 99–120. https://doi.org/10.1177/014920639101700108

Bienhaus, F., & Haddud, A. (2018). Procurement 4.0: Factors influencing the digitisation of procurement and supply chains. *Business Process Management Journal*, *24*(4), 965–984. https://doi.org/10.1108/BPMJ-06-2017-0139

Gottge, S., Menzel, T., & Forslund, H. (2020). Industry 4.0 technologies in the purchasing process. *Industrial Management & Data Systems*, *120*(4), 730–748. https://doi.org/10.1108/IMDS-05-2019-0304

Gupta, R. K., Belkadi, F., Buergy, C., Bitte, F., da Cunha, C., Buergin, J., Lanza, G., & Bernard, A. (2018). Gathering, evaluating and managing customer feedback during aircraft production. *Computers & Industrial Engineering*, *115*, 559–572. https://doi.org/10.1016/j.cie.2017.12.012

Leung, E. K. H., Lee, C. K. H., & Ouyang, Z. (2022). From traditional warehouses to physical internet hubs: A digital twin-based inbound synchronization framework for PI-order management. *International Journal of Production Economics*, *244*, 108353. https://doi.org/10.1016/j.ijpe.2021.108353

Meindl, B., Ayala, N. F., Mendonça, J., & Frank, A. G. (2021). The four smarts of Industry 4.0: Evolution of ten years of research and future perspectives. *Technological Forecasting and Social Change*, *168*, 120784. https://doi.org/10.1016/j.techfore.2021.120784

Montecchi, M., Plangger, K., & Etter, M. (2019). It's real, trust me! Establishing supply chain provenance using blockchain. *Business Horizons*, *62*(3), 283–293. https://doi.org/10.1016/j.bushor.2019.01.008

Mujahid Ghouri, A., Mani, V., Jiao, Z., Venkatesh, V. G., Shi, Y., & Kamble, S. S. (2021). An empirical study of real-time information-receiving using Industry 4.0 technologies in downstream operations. *Technological Forecasting and Social Change*, *165*, 120551. https://doi.org/10.1016/j.techfore.2020.120551

Queiroz, M. M., Fosso Wamba, S., Machado, M. C., & Telles, R. (2020). Smart production systems drivers for business process management improvement. *Business Process Management Journal*, *26*(5), 1075–1092. https://doi.org/10.1108/BPMJ-03-2019-0134

Queiroz, M. M., Pereira, S. C. F., Telles, R., & Machado, M. C. (2021). Industry 4.0 and digital supply chain capabilities. *Benchmarking: An International Journal*, *28*(5), 1761–1782. https://doi.org/10.1108/BIJ-12-2018-0435

Wamba, S. F., & Queiroz, M. M. (2020). Blockchain in the operations and supply chain management: Benefits, challenges and future research opportunities. *International Journal of Information Management*, *52*, 102064. https://doi.org/10.1016/j.ijinfomgt.2019.102064

Wernerfelt, B. (1984). A resource-based view of the firm. *Strategic Management Journal*, *5*(2), 171–180. https://doi.org/10.1002/smj.4250050207

Winkelhaus, S., Grosse, E. H., & Morana, S. (2021). Towards a conceptualisation of order Picking 4.0. *Computers & Industrial Engineering*, *159*, 107511. https://doi.org/10.1016/j.cie.2021.107511

Yan, J., Liu, J., & Tseng, F.-M. (2020). An evaluation system based on the self-organizing system framework of smart cities: A case study of smart transportation systems in China. *Technological Forecasting and Social Change*, *153*, 119371. https://doi.org/10.1016/j.techfore.2018.07.009

Digital Supply Structure and Configuration

7

7.1 DIMENSIONS OF THE DIGITAL SUPPLY CHAIN IMPLEMENTATION

The implementation of digital supply chain projects requires comprehensive efforts from not only the organization in which the project is being implemented but also the many different partners that are involved. To present the dynamics of the digital supply chain implementation, two aspects require particular attention: structure and configuration. While structure refers to a set of internal and external tangible and intangible resources (people with technical and communication skills; management leadership; technologies acquired, financial flow, etc.), configuration refers to the management and integration of different activities both internally and throughout the supply chains.

Table 7.1 shows the main dimensions of digital transformation implementation. Five dimensions are considered the foundations of implementation: operational flow remodeling, management process, supply chain process remodeling, integration modeling process throughout the supply chain, and digital supply chain value. These dimensions also require interaction with the implementation of the technologies. In the next subsection, these dimensions will be detailed.

DOI: 10.1201/9781003226468-7

TABLE 7.1   Main dimensions of digital transformation implementation

DIMENSION	BRIEF DESCRIPTION
Operational flow remodeling	Refers to the remodeling of the main logistics and supply chain activities flows, including inbound and outbound (i.e., raw materials flow, information, money, finished products, storage, picking, packing, transportation, reverse flow, etc.)
Management process	Refers to the leadership of the directors and managers from the organization in which the project is being implemented in order to ensure the necessary resources for the project, defining the goals, plan, and control
Supply chain process remodeling	Refers to the identification, mapping, and remodeling of the main processes which require a supply chain member (i.e., procurement, demand management, inventory control, etc.)
Integration modeling process throughout the supply chain	Refers to the information sharing between the organization and its supply chain members in order to integrate the efforts and collaboration into the process smoothly (i.e., costs, lead times, resources, constraints, etc.)
Digital supply chain value	Refers to the value creation enabled by the digital transformation (i.e., responsiveness improvement, mass customization, customer experience, risk exposure minimization, transparency, cost reduction, innovative products/services, etc.)

7.2 THE ROLE OF THE DIGITAL SUPPLY CHAIN STRUCTURE AND CONFIGURATION

The dimensions of the digital transformation implementation follow an integrated and sequential flow as shown in Figure 7.1. The implementation process starts with the operational flow remodeling toward integrating the process into the supply chains. It is important to note that, despite the sequential perspective, all dimensions interact with each other.

Operational flow remodeling: This dimension focuses on the basic logistics and supply chain activities (L&SC; Garay-Rondero et al., 2019). The main inbound and outbound logistics, as outlined in Table 7.1, require

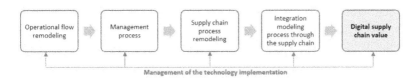

FIGURE 7.1   Sequential flow of the dimensions of the digital transformation implementation.

in-depth rounds of remodeling to determine the best flow. In addition, as shown in Figure 7.1, this and all the dimensions interact with the management of the technology's implementation. For instance, in a flow related to the raw materials, the new (remodeled) flow should consider the best technology or a set of technologies that can support the process. In the same vein, IoT combined with augmented reality can add more value to this type of flow. Examples of the value of operational flows include time reduction, quality and confidence improvement in the process, the integrity of the materials and workers, etc. This reasoning should be employed for each individual activity.

Management process: In the management process, the leadership abilities of the decision-makers, especially directors and managers, must be effective to acquire the different tangible, intangible, and human resources necessary for each stage of the project. These resources require good planning and control to be used in line with the TOP framework.

Supply chain process remodeling: In this dimension, supply chain remodeling involves both operational and strategic levels of the organization's supply chain. As such, the identification and mapping of the main process through the supply chains with which the organization has a relationship plays a critical role in remodeling the process. However, completely remodeling a supply chain process will require considerable collaboration among the members. Otherwise, the process will work only partially or not at all.

Integration modeling process throughout the supply chain: This is a sequence of the previous dimensions. For this dimension, integration modeling requires both operational and strategic level approaches to the organization's supply chain. Thus, one of the critical aspects is the integration of the supply chain, mainly considering the strategies related to information sharing between the members, enabled by cutting-edge technologies (Büyüközkan & Göçer, 2018). For example, in the demand management process, as it applies specifically to the hospitality sector (tourism management), organizations can use big data analytics approaches to gain a more in-depth understanding and novel insights to support the innovative services (Shamim et al., 2021).

Digital supply chain value: Digital supply chain value is created when the previous dimensions successfully interact with each other with the support of cutting-edge technologies. In terms of value for customers, the successful implementation of key technologies supported by the TOP view can significantly improve the customer experience. For instance, visibility, transparency, and customer trust with regard to products can be leveraged using blockchain applications throughout the supply chains (Sunny et al., 2020).

From the supplier's side, technologies like IoT, AI, BDA, sensors, RFID, etc. can improve integration, collaboration, and information sharing in real time with the main network partners (Bag et al., 2020). This means that companies can become more agile and reduce transaction costs.

From the firm's side, enhanced customer integration can support the development of new products and services according to real customer demand (Jain et al., 2021; Li et al., 2021), thus enabling a new value experience for customers. Additionally, customer interaction, offline or online, can allow customers to actively participate in the firm's co-creation process.

7.2.1 Technology Implementation

Before a company can implement one or more technologies, it must go through two prior stages. Organizations must be willing to adopt (intention to adopt) and accept the technology (adoption). In this regard, **intention to adopt** refers to the knowledge accumulation about the characteristics of the technology; **adoption** refers to the acceptance of the technology based on its benefits and the organization's approval, and **implementation** refers to initial use by converting technology planning into action.

Taking into account the implementation stage requires significant efforts, strong concepts, and reliable techniques from business process management (Fischer et al., 2020) and project management (Garay-Rondero et al., 2019). For example, aspects related to transparent communication between the stakeholders include governance, support from top management, coordination and integration of process architecture, training, knowledge sharing, IT support, collaboration support for stakeholders, standards, controls, and patterns of the process modeling, an atmosphere that enables the stakeholders' commitment, etc. (Fischer et al., 2020).

For implementation, the organization needs to set a single direction for employees and all other stakeholders. The plan should be shored up by the communication strategies, which in turn support the integration of all layers and commitment. It is important to note that the implementation stage requires adequate time to adapt and integrate the process, organizational structures, and supply chains (Reuschl et al., 2022). Otherwise, negative effects will appear in the project.

7.3 ASSESSMENT OF DIGITAL SUPPLY CHAIN PERFORMANCE

To analyze whether the implementation of the digital supply chain project (or certain modules) is successful, key performance indicators (KPIs) must be used (Oliveira-Dias et al., 2022; Rasool et al., 2022). From this perspective, the performance categories and KPIs shown in Table 7.2 could be used to assess the outcome of the digital transformation project.

Table 7.2 summarizes the categories, KPIs, outcomes, and the main approaches/technologies. The categories are the organization, top

TABLE 7.2 Categories of the digital supply chain performance

CATEGORY	KPI	OUTCOME	APPROACHES AND/ OR TECHNOLOGIES
Organization	Digital culture value dissemination	Improvement of people reached with a clear vision of values of the digital transformation	Internal and external communication systems
	Support for digital transformation initiatives	Investments in technologies, people, and the organization's infrastructure	Incorporation of a set of key cutting-edge technologies
Top management support	Leadership of workers	Productivity of the workers in digital transformation-related activities	Participation of the workers in the implementation of the project
	Resource acquisition	Percentage of the acquired resources as the total number needed by the organization	Procurement 4.0, ERP, cloud computing
	Capabilities enabled by resources	Number of capabilities supported by the resources acquired	Development of key capabilities enabled by a set of key cutting-edge technologies
	Project management deliverables	Deliveries of the digital initiatives (projects) made on time	Technologies related to project management, cloud computing, ERP, digital twin

(Continued)

TABLE 7.2   (Continued)

CATEGORY	KPI	OUTCOME	APPROACHES AND/ OR TECHNOLOGIES
Workers	Digital literacy skills, including technical and non-technical	Development of different skills (statistical, math, programming, communication, storytelling, technology implementation, etc.)	AI, BDA, digital twin, simulation, dashboards
	Number of hours during a year dedicated to training	Engagement with training related to digital transformation initiatives	New technical and non-technical skills developments
	Participation in products or services improved or created	Innovation enabled by digital transformation skills	AI, BDA, simulation, IoT, VR, AR
Production	Quality improvement	Waste, re-work and scrap minimization	Simulation and optimization tools
	Scheduling processes	Maximization of resource use (machinery, workers, facilities)	Simulation and optimization tools
	Production processes	Cycle time minimization	Simulation, optimization, AM/3D printing
	Production flexibility	Number of unscheduled orders fulfilled	Simulation, optimization, AI, BDA
Warehouse	Receiving productivity and accuracy	The volume of goods/ materials received and accurately checked	WMS, sensors, AR, IoT, RFID, QR code
	Storage and inventory management	Productivity and accuracy of the products stocked	WMS, sensors, AR, VR, IoT, RFID, QR code
	Picking processing	Order picking accuracy and productivity improvement	WMS, sensors, AR, VR, IoT, RFID, QR code, cobots, vision picking, picking by light, picking by voice, automated guided vehicles (AGVs), drones
	Loaded trucks	Improvement of the number of trucks loaded	AGVs, cobots

Transportation	Load fulfillment optimization	Improvement of the truck capacity utilization	Simulation, AI, BDA
	Vehicle time utilization	The time that the vehicle is used as a percentage of the scheduled time	Simulation, optimization, AI, BDA
	Fuel cost minimization	Fuel costs per km	Blockchain, IoT, GPS
	On-time pickup and delivery	Percentage of the pickups made on time	Blockchain, IoT, GPS
	Minimization of CO_2 emissions	The number of GHGs emitted per mile tripped	Blockchain, IoT, GPS
Customers	On-time delivery	Percentage of the orders delivered according to the customer time	Blockchain, IoT, GPS
	Tracing and tracking	Visibility of the product journey	Blockchain, IoT, GPS
	% of correct orders received	Percentage of the orders delivered according to customer needs	Blockchain, IoT, GPS
	% of stockout	Minimization of stockouts	AI, BDA, simulation, IoT, RFID, QR code
	Co-creation participation	Customer experience improvement	CRM, ERP, AI, BDA, NFC, Bluetooth, IoT
Suppliers	Integration	Minimization of information asymmetry	Digital twin, blockchain, AI
	Collaboration	Resource and capability sharing	Digital twin, blockchain, AI, cloud computing
	Knowledge sharing	Incorporation of some of the partners' best practices	Digital twin, blockchain, AI, cloud computing, ERP
	Lead time minimization	Improvement in the speed of supply	Digital twin, blockchain, AI, cloud computing, ERP
	Flexibility	Improvement of the ability to absorb unscheduled orders	Digital twin, blockchain, AI, cloud computing, ERP

management support, production, warehouse, transportation, customers, and suppliers. Some of the categories are related to a strategic view and do not refer to a particular technology; thus, the outcome is more qualitative and strategic. Other categories are connected with technologies and, as such, refer to a quantitative outcome.

7.4 SHORT CASE STUDY: IMPLEMENTATION OF A DIGITAL SUPPLY CHAIN PROJECT

The Omega Company is a global furniture retailer. Omega's e-commerce sales have increased significantly and now correspond to 80% of the company's revenue. Due to the COVID-19 pandemic, more people have started working full or part time from home. As a result, aspects related to product delivery, agility, flexibility, and customization have generated several complaints from customers. Omega only has an e-commerce platform, and so the L&SC Director has suggested a project to create a digital supply chain to better address these issues. How could the company use the information from Figure 7.1 (Sequential flow of the dimensions of the digital transformation implementation) in their L&SC operations to support a project on supply chain structure and configuration?

REFERENCES

Bag, S., Wood, L. C., Mangla, S. K., & Luthra, S. (2020). Procurement 4.0 and its implications on business process performance in a circular economy. *Resources, Conservation and Recycling*, *152*, 104502. https://doi.org/10.1016/j.resconrec.2019.104502

Büyüközkan, G., & Göçer, F. (2018). Digital supply chain: Literature review and a proposed framework for future research. *Computers in Industry*, *97*, 157–177. https://doi.org/10.1016/j.compind.2018.02.010

Fischer, M., Imgrund, F., Janiesch, C., & Winkelmann, A. (2020). Strategy archetypes for digital transformation: Defining meta objectives using business process management. *Information and Management*, *57*(5). https://doi.org/10.1016/j.im.2019.103262

Garay-Rondero, C. L., Martinez-Flores, J. L., Smith, N. R., Caballero Morales, S. O., & Aldrette-Malacara, A. (2019). Digital supply chain model in Industry 4.0. *Journal of Manufacturing Technology Management*, *31*(5), 887–933. https://doi.org/10.1108/JMTM-08-2018-0280

Jain, G., Paul, J., & Shrivastava, A. (2021). Hyper-personalization, co-creation, digital clienteling and transformation. *Journal of Business Research*, *124*, 12–23. https://doi.org/10.1016/j.jbusres.2020.11.034

Li, S., Peng, G., Xing, F., Zhang, J., & Zhang, B. (2021). Value co-creation in industrial AI: The interactive role of B2B supplier, customer and technology provider. *Industrial Marketing Management*, *98*, 105–114. https://doi.org/10.1016/j.indmarman.2021.07.015

Oliveira-Dias, D., Maqueira-Marín, J. M., & Moyano-Fuentes, J. (2022). The link between information and digital technologies of Industry 4.0 and agile supply chain: Mapping current research and establishing new research avenues. *Computers and Industrial Engineering*, *167*, 108000. https://doi.org/10.1016/j.cie.2022.108000

Rasool, F., Greco, M., & Grimaldi, M. (2022). Digital supply chain performance metrics: A literature review. *Measuring Business Excellence*, 26(1), 23–38. https://doi.org/10.1108/MBE-11-2020-0147

Reuschl, A. J., Deist, M. K., & Maalaoui, A. (2022). Digital transformation during a pandemic: Stretching the organizational elasticity. *Journal of Business Research*, *144*, 1320–1332. https://doi.org/10.1016/j.jbusres.2022.01.088

Shamim, S., Yang, Y., Zia, N. U., & Shah, M. H. (2021). Big data management capabilities in the hospitality sector: Service innovation and customer generated online quality ratings. *Computers in Human Behavior*, *121*, 106777. https://doi.org/10.1016/j.chb.2021.106777

Sunny, J., Undralla, N., & Madhusudanan Pillai, V. (2020). Supply chain transparency through blockchain-based traceability: An overview with demonstration. *Computers and Industrial Engineering*, *150*, 106895. https://doi.org/10.1016/j.cie.2020.106895

Operations Management, Process, and Business Model Reconfiguration

8

8.1 OPERATIONS MANAGEMENT IN THE DIGITAL ERA

Operations management (OM) is a field responsible for managing all processes required to produce goods and services (Silver, 2004). In short, OM is a strategic area that must integrate and manage different inputs that go into many processes in order to transform them into outputs (goods/services). In this context, the transformation process requires different types of inputs, such as people, materials, facilities, machinery, and technology. These inputs need to be managed and orchestrated carefully to meet demand according to the customer's requirements. Figure 8.1 shows OM as a strategic field where inputs are converted into goods.

The transformation process is a complex task to be managed because, in most organizations, it involves multiple processes and activities, which in

DOI: 10.1201/9781003226468-8

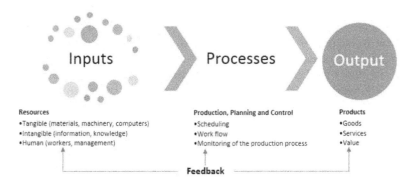

FIGURE 8.1   Operations management and the transformation process.

turn require different types of resources that should be integrated so as to meet demand efficiently. Most inputs are highly dependent on the environment and thus make OM more complex. In recent years, with the advancement of Industry 4.0 and the digital transformation phenomenon, the OM field has experienced major changes by integrating cutting-edge technologies, knowledge, and skills. The integration of Industry 4.0 and digital transformation has allowed OM to shift to OM 4.0 (described in greater detail in the next subsection), which operates more efficiently and is thus able to create greater value for products and services (Choi et al., 2022; Olsen & Tomlin, 2020).

8.2 OPERATIONS MANAGEMENT 4.0

This book defines OM 4.0 as the integration of Industry 4.0 and digital transformation paradigms associated with the planning, designing, management, and control of all processes to produce goods and services, with high levels of interoperability, interconnectivity, and hyperconnection throughout supply chains, in order to maximize the value offered to stakeholders. OM 4.0 is a still nascent field with the potential to remodel several established business models at the organizational and supply chain levels.

In the Industry 4.0/OM 4.0 views, the integration between different types of elements (e.g., machinery, objects, and computers), mainly through sensors, IoT, and CPS, can support a new generation of processes and operations. Some can be performed with high levels of intelligence, while with certain approaches, autonomous decisions are possible (Qin et al., 2016). Furthermore, from the standpoint of interoperability, interconnectivity, and hyperconnection, supply chains can be completely digitalized, enabling a genuine value chain that integrates people, machines, and systems. One recent technology that makes this possible is the digital twin (Ivanov & Dolgui, 2021; Leung et al., 2022).

Another significant feature that has emerged in the Industry 4.0 era and which is impacting OM 4.0 is the smart approaches (i.e., smart products, smart manufacturing, smart working, smart supply chains, smart services) (Meindl et al., 2021; Queiroz et al., 2020). For instance, smart products basically use sensors to record data about their journey in the supply chains as well as feedback from customers about their experiences. Smart manufacturing, also known as the smart factory, is a new generation of manufacturing, supported mainly by cutting-edge technologies like CPS, BDA, and IoT, among others. The smart factory is able to operate autonomously, supported by a technological network, by employing these and other technologies (Lee, 2015).

The smart supply chain refers to the interconnection of the network members by applying cutting-edge technologies using a smart approach, providing, in most cases, real-time information, collaboration, responsiveness, and transparency between the partners across all stages of the network (Akyuz & Gursoy, 2019; Wu et al., 2016).

Smart products are embedded with different sensors that enable storage and information sharing through the supply chains, giving high levels of feedback to network members, including factories, transport companies, retailers, etc. Using smart products implies having an optimized network that minimizes waste in redundant processes and provides more transparent and real-time information (Qin et al., 2016). In this way, smart products are able to make decisions in an autonomous way, supported by AI applications (Büyüközkan & Göçer, 2018; Hahn, 2020).

Smart working is related to the integration between humans and technologies to support production system processes in a flexible and optimized manner (Dornelles et al., 2022; Frank et al., 2019). In the context of OM in digital supply chains, smart working can empower workers to create and reconfigure supply chains supported by technologies like AI, digital twin, BDA, blockchain, etc. and, consequently, contribute to adding more value to the supply chains. Smart working not only leverages the performance of organizations and supply chains, but it also enhances worker satisfaction.

Finally, smart services refer to the use of cutting-edge technologies in the processes and operations through the supply chains (Ardolino et al., 2018). Such services can include BDA, AI, cloud applications, blockchain, and digital twin to store transactions, analyze data, develop predictive models, make forecasts, perform tracing and tracking, monitor disruptions, improve customer service, and more.

8.3 PROCESS AND BUSINESS MODEL RECONFIGURATION

The OM field, with the digital transformation approaches, is supporting the remodeling of processes and the entire business model. Table 8.1 shows the main benefits of some key reconfigurations, including labor, process automatization, decentralization, design/prototype, transactions, control, monitoring, suppliers, and customers.

For instance, production system configurations, which are highly intensive in terms of operational labor and machines, are now experiencing a

TABLE 8.1   Examples and main benefits of process reconfiguration

EXAMPLE OF RECONFIGURATIONS	MAIN TECHNOLOGIES TO SUPPORT	EXAMPLE OF BENEFITS
Labor	BDA, AI, simulation	Accidents reduction, efficiency in the process, enhancement of the analysis and monitoring, more participation in strategic issues
Process automation	AGVs, CPS, IoT	Economies of scale in production, greater efficiency, less repetitive work
Decentralization	AM/3D printing	Minimization of the dependency on transportation of goods, responsiveness to meet the demand
Design/prototype	Digital twin, AM/3D printing	Speed up time to market
Transactions	Blockchain	Costs minimization, traceability, accountability improvement
Control	IoT, RFID, AI	Agility and accuracy in the inventory
Monitoring	Digital twin	Anticipation of environmental disruptions with real-time monitoring
Suppliers	Blockchain, digital twin, AI, BDA, AM	Efficiency in attending the purchase orders, visibility and monitoring improvement, forecast accuracy
Customers	Social media, AI, IoT	Online and offline feedback, co-participation in the production

"revolution" by reconfiguring the traditional process with CPS, IoT, AGVs, AI, digital twin, etc. In addition, the old perspective of human labor, which focused on repetitive tasks, is shifting toward more analytical work where only a few employees are needed to control and monitor the entire production system instead of just one small unit (Dornelles et al., 2022).

Furthermore, with the emergence of cloud computing approaches, several maintenance processes can be performed online by experts and a specialized workforce from any country. This means that digital transformation can decentralize several production system tasks (Rui et al., 2022). Another significant example of the process and business model reconfiguration is the use of additive manufacturing/3D printing to decentralize production. With AM/3D printing, companies can operate with small local partners to produce close to their market, thus gaining in agility and minimizing costs (Bogers et al., 2016). Regarding the design and prototype process, AM/3D printing is able to speed up these activities at a much lower cost than traditional approaches.

When it comes to transactions involved in the internal or external process, with blockchain technologies companies can drastically reduce transaction costs, as well as make the process more reliable and offer high levels of traceability of the goods throughout the supply chains (Wamba & Queiroz, 2022). From the control perspective, technologies like IoT, RFID, and AI can reconfigure processes related to managing inventory, and with a digital twin, monitoring processes can be made in real time.

By employing technologies like blockchain, digital twin, AI, BDA, and AM, OM processes with suppliers can be reconfigured, mainly in processes related to purchasing orders, forecasting, monitoring, transaction costs, etc. Finally, with regard to representative reconfiguration in business processes with customers, technologies like social media, AI, and IoT can provide closer interactions with customers as well as online and offline feedback about product use (Airani & Karande, 2022; Paiola et al., 2021).

8.4 SHORT CASE STUDY: OPERATIONS MANAGEMENT 4.0 TO RECONFIGURE CRITICAL PROCESSES

A logistics service provider is facing some huge challenges in its management of processes related to booked cargo and integration with customers and suppliers. The company operates globally and has recently seen its market share fall significantly. Some of the main complaints reported in the last year include freight costs that do not appear to be competitive and an increase in problems related to overbooking, incorrect information in upstream and downstream processes, and follow-up. The company needs to reconfigure these and other critical processes by integrating strategies from the OM 4.0 approach. In order to support the company, what plan would you recommend the decision-makers implement?

REFERENCES

Airani, R., & Karande, K. (2022). How social media effects shape sentiments along the twitter journey? A Bayesian network approach. *Journal of Business Research*, *142*, 988–997. https://doi.org/10.1016/j.jbusres.2021.12.071

Akyuz, G. A., & Gursoy, G. (2019). *Becoming Smart, Innovative, and Socially Responsible in Supply Chain Collaboration* (pp. 919–941). https://doi.org/10.4018/978-1-5225-7362-3.ch069

Ardolino, M., Rapaccini, M., Saccani, N., Gaiardelli, P., Crespi, G., & Ruggeri, C. (2018). The role of digital technologies for the service transformation of industrial companies. *International Journal of Production Research*, *56*(6), 2116–2132. https://doi.org/10.1080/00207543.2017.1324224

Bogers, M., Hadar, R., & Bilberg, A. (2016). Additive manufacturing for consumer-centric business models: Implications for supply chains in consumer goods manufacturing. *Technological Forecasting and Social Change*, *102*, 225–239. https://doi.org/10.1016/j.techfore.2015.07.024

Büyüközkan, G., & Göçer, F. (2018). Digital supply chain: Literature review and a proposed framework for future research. *Computers in Industry*, *97*, 157–177. https://doi.org/10.1016/j.compind.2018.02.010

Choi, T., Kumar, S., Yue, X., & Chan, H. (2022). Disruptive technologies and operations management in the Industry 4.0 era and beyond. *Production and Operations Management*, *31*(1), 9–31. https://doi.org/10.1111/poms.13622

Dornelles, J. de A., Ayala, N. F., & Frank, A. G. (2022). Smart working in Industry 4.0: How digital technologies enhance manufacturing workers' activities. *Computers & Industrial Engineering*, *163*, 107804. https://doi.org/10.1016/j.cie.2021.107804

Frank, A. G., Dalenogare, L. S., & Ayala, N. F. (2019). Industry 4.0 technologies: Implementation patterns in manufacturing companies. *International Journal of Production Economics*, *210*, 15–26. https://doi.org/10.1016/j.ijpe.2019.01.004

Hahn, G. J. (2020). Industry 4.0: A supply chain innovation perspective. *International Journal of Production Research*, *58*(5), 1425–1441. https://doi.org/10.1080/00207543.2019.1641642

Ivanov, D., & Dolgui, A. (2021). A digital supply chain twin for managing the disruption risks and resilience in the era of Industry 4.0. *Production Planning & Control*, *32*(9), 775–788. https://doi.org/10.1080/09537287.2020.1768450

Lee, J. (2015). Smart factory systems. *Informatik-Spektrum*, *38*(3), 230–235. https://doi.org/10.1007/s00287-015-0891-z

Leung, E. K. H., Lee, C. K. H., & Ouyang, Z. (2022). From traditional warehouses to physical internet hubs: A digital twin-based inbound synchronization framework for PI-order management. *International Journal of Production Economics*, *244*, 108353. https://doi.org/10.1016/j.ijpe.2021.108353

Meindl, B., Ayala, N. F., Mendonça, J., & Frank, A. G. (2021). The four smarts of Industry 4.0: Evolution of ten years of research and future perspectives. *Technological Forecasting and Social Change*, *168*, 120784. https://doi.org/10.1016/j.techfore.2021.120784

Olsen, T. L., & Tomlin, B. (2020). Industry 4.0: Opportunities and challenges for operations management. *Manufacturing and Service Operations Management*, *22*(1), 113–122. https://doi.org/10.1287/msom.2019.0796

Paiola, M., Schiavone, F., Grandinetti, R., & Chen, J. (2021). Digital servitization and sustainability through networking: Some evidences from IoT-based business models. *Journal of Business Research*, *132*, 507–516. https://doi.org/10.1016/j.jbusres.2021.04.047

Qin, J., Liu, Y., & Grosvenor, R. (2016). A categorical framework of manufacturing for Industry 4.0 and beyond. *Procedia CIRP*, *52*, 173–178. https://doi.org/10.1016/j.procir.2016.08.005

Queiroz, M. M., Fosso Wamba, S., Machado, M. C., & Telles, R. (2020). Smart production systems drivers for business process management improvement. *Business Process Management Journal*, *26*(5), 1075–1092. https://doi.org/10.1108/BPMJ-03-2019-0134

Rui, L., Yang, S., Gao, Z., Li, W., Qiu, X., & Meng, L. (2022). Smart network maintenance in edge cloud computing environment: An allocation mechanism based on comprehensive reputation and regional prediction model. *Journal of Network and Computer Applications*, *198*, 103298. https://doi.org/10.1016/j.jnca.2021.103298

Silver, E. A. (2004). Process management instead of operations management. *Manufacturing and Service Operations Management*, 6(4), 273–279. https://doi.org/10.1287/msom.1040.0055

Wamba, S. F., & Queiroz, M. M. (2022). Industry 4.0 and the supply chain digitalisation: A blockchain diffusion perspective. *Production Planning & Control*, *33*(2–3), 193–210. https://doi.org/10.1080/09537287.2020.1810756

Wu, L., Yue, X., Jin, A., & Yen, D. C. (2016). Smart supply chain management: A review and implications for future research. *The International Journal of Logistics Management*, *27*(2), 395–417. https://doi.org/10.1108/IJLM-02-2014-0035

Digital Transformation and Operations Research Models

9.1 OPERATIONS RESEARCH: A BASIC APPROACH

According to the IFORS (2022), operations research (OR) is a field that

> encompasses the development and the application of a wide range of problem-solving methods and techniques applied in the pursuit of improved decision-making and efficiency, such as mathematical optimization, simulation, queueing theory and other stochastic models. The OR methods and techniques involve the construction of mathematical models that aim at describing a problem.

Mathematical models are developed based on four main steps. First, the problem to be considered by the model must be formalized (**Step 1, Conceptualization**). Next, the model is developed with the variable specifications (**Step 2, Modeling**). The model then begins the solving process using a formal approach (**Step 3, Model solving**). Finally, based on the feedback from the outputs, the results are implemented (**Step 4, Implementation**), supported by the model solving (Mitroff et al., 1974). It is important to note that all steps are connected to each other. In addition, in the relationship between the conceptual model and the solution, there is feedback, while validation must occur between the problem situation and scientific model. Figure 9.1 highlights these steps. With the digital transformation phenomenon, the interaction of the steps from Figure 9.1 is more dynamic and, in several cases, happens in real time.

It is important to note that OR is considered a subset of operations management (OM), focused on exploring real-world problems with mathematical and related approaches. The aim is to provide decision-makers with solutions, mainly applying queueing theory, dynamic programming, decision theory, game theory, stochastic models, combinatorial optimization, etc. (Bertrand & Fransoo, 2002).

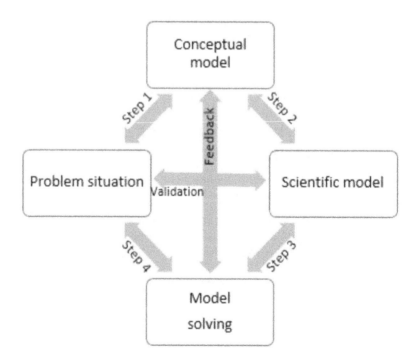

FIGURE 9.1   Problem-solving flow in OR. Adapted from Mitroff et al. (1974).

9.2 OPERATIONS RESEARCH 4.0

With the exponential advances made by Industry 4.0 and the digital transformation phenomena, the OR field also has benefited and integrated some of the main best practices and technologies. This book defines OR 4.0 as a field that uses cutting-edge technologies such as artificial intelligence (AI), big data, and algorithms in real-world problems to support the development, modeling, solving, and implementation of the solution by employing the finest know-how from different but complementary areas like computer science, optimization, stochastic processes, game theory, decision science, graph theory, probability, and simulation, among others.

Accordingly, OR 4.0 is able to develop different types of solutions in a short time and at low cost with high levels of quality, and in several cases, can provide optimal solutions. One of the main approaches in OR/OR 4.0 is the mathematical programming approach. Techniques related to linear programming, integer programming, and mixed-integer linear programming are classical and popular techniques to develop models. In addition, with the exponential growth of the processing capacity of computers, algorithm optimization—including exact methods, heuristics, and metaheuristics—has evolved to another level, using robust commercial packages of solvers (i.e., Gurobi, IBM ILOG CPLEX, etc.).

9.3 SOME CONSIDERATIONS ON ARTIFICIAL INTELLIGENCE TECHNOLOGIES

AI technologies have been used in a vast range of contexts. However, there is some confusion in the literature about what AI is, the main types of AI, and some applications using AI. In order to shed more light on this cutting-edge technology, we present a simple framework for the organization of AI tools.

Table 9.1 highlights some popular AI tools. Thus, machine learning (ML), deep learning (DL), and natural language processing (NLP) are three of the most popular AI tools/approaches. Furthermore, other AI applications and approaches, like expert systems (ES), which has a long tradition in a number of business contexts, and recent AI applications such as robotics, speech recognition, and vision applications, are gaining momentum among practitioners and scholars (Collins et al., 2021).

TABLE 9.1 Popular AI tools and methods

AI APPROACH	AI TECHNOLOGY	BRIEF EXPLANATION	RELATED LITERATURE
Learning technologies (LT)	Machine learning (ML) Deep learning (DL)	AI techniques that employ previous experience based on patterns and training	(Kang et al., 2020; Sabahi & Parast, 2020; Smiti & Soui, 2020)
Network applications (NA)	Artificial neural networks (ANN) Convolutional neural network (CNN) Bayesian Network (BN)	AI techniques that simulate the brain experience in processing data	(Hosseini & Ivanov, 2020; Lima-Junior & Carpinetti, 2019; Wang et al., 2019)
Natural language processing (NLP)	Speech recognition (SR) Optical Character Recognition (OCR) Sentiment analysis (SA) Machine translation (MT) Chatbots (CB)	AI techniques for speech recognition, question answering, language production, translation, etc.	(Cebollada et al., 2020; Farhan et al., 2020; Paschen et al., 2019; Sun & Medaglia, 2019; Tellez et al., 2017)
Automation technology (AT)	Robotic process automation (RPA)	AI techniques that mimic human manual tasks for business process automation	(Syed et al., 2020)
Expert systems (ES)	Fuzzy systems (FS) Rule-based (RB) Real-time expert systems (RTES) Backward and forward chaining (BFC)	AI techniques for simulating the decision-making process, supported by an in-depth knowledge set and rules	(Caiado et al., 2021; del Mar Roldán-García et al., 2018; Kim et al., 2008; Lee et al., 2018)
Computer vision (CV)	Object recognition (OBR) Image processing (IP) Event detection and tracking (EDT)	AI techniques that interpret, understand, and simulate the real-world visually	(Cebollada et al., 2020; De Oliveira et al., 2016; Klaib et al., 2021)

| Algorithms approach (AA) | Heuristics and Metaheuristics (Tabu search, Genetic algorithm, Simulated annealing, Swarm intelligence, Iterated Local Search, Hill Climbing, Ant Colony Optimization, Evolutionary algorithms) | AI techniques that employ powerful algorithms from an optimization perspective. Heuristics cannot guarantee the optimal solution but are able to achieve them in several cases. | (Dulebenets, 2019; Elshaer & Awad, 2020; Silva et al., 2021; Torres-Jiménez & Pavón, 2014) |

9.4 OPERATIONS RESEARCH MODELS IN THE AGE OF DIGITAL TRANSFORMATION

In order to provide a holistic view, Table 9.2 summarizes some of the main approaches related to the OR 4.0 view, including algorithm approaches and cutting-edge technologies such as AI and big data analytics.

TABLE 9.2 Examples of OR 4.0 operations, tools/techniques, and their benefits

TYPE OF OPERATION/ PROCESS/ACTIVITY	TOOLS/TECHNIQUES	EXAMPLES OF BENEFITS
Network modeling optimization (supply chain)	Simulation, optimization (heuristics/ metaheuristics, exact methods)	Configuration/reconfiguration of the network
Scheduling	Simulation, optimization (heuristics/ metaheuristics, exact methods)	Selection of the best resources (human, machines, time) to meet the sequence of the orders
Inventory management	AI, big data analytics, multiple-criteria decision-making (MCDM)	Real-time monitoring and control; leveraging of the accuracy levels

(Continued)

TABLE 9.2   (Continued)

TYPE OF OPERATION/ PROCESS/ACTIVITY	TOOLS/TECHNIQUES	EXAMPLES OF BENEFITS
Resource allocation	Optimization (heuristics/ metaheuristics, exact methods), MCDM	Dynamic resource allocation according to the demand pattern (i.e., hospital bed allocation)
Facility location	Simulation, optimization (heuristics/ metaheuristics, exact methods), MCDM	Support for choosing the best geographic location for the company's operations considering the environmental dynamics
Transportation and routing	Simulation, optimization (heuristics/ metaheuristics, exact methods)	Faster, more efficient, and more cost-effective pickup and delivery operations
Layout design	Simulation, optimization (heuristics/ metaheuristics, exact methods)	Improvement of the physical placement, production efficiency, and quality
Demand forecasting	AI, big data analytics, MCDM	Robust models to support the demand prediction considering many variables from different sources
Procurement	Cloud-based, game theory, process automation, AI, blockchain, MCDM	Agility, reliability, and accountability in the purchasing process through the network
Project management	Digital twin, AI, big data analytics, MCDM	Improvement of the control and monitoring of the scope, budget, and team; optimization of resource allocation
Risk management	Digital twin, AI, big data analytics, blockchain, MCDM	Real-time monitoring of the supply chain and minimization of uncertainties

9.5 SHORT CASE STUDY: MANAGING AND OPTIMIZING THE HTAA HOSPITAL'S OPERATIONS

HTAA is a hospital with global operations. Recently, the hospital's directors, including the OM and supply chain management (SCM) directors, were involved in implementing a global digital transformation project. Neither the OM nor the SCM director was an expert in the digital transformation phenomenon. The basic scope reported by the directors related to the digitization approach, that is, the remodeling of the physical activities (essentially paper-based tasks) into a digital version. Due to the COVID-19 pandemic, the HTAA hospital is facing several challenges at its global sites. For instance, the surgery department does not have enough surgeons or intensive care resources, and the inventory of various types of equipment and drugs has dropped to unprecedented levels, causing some units to close. How can OR 4.0 support the directors in improving the HTAA hospital's operations?

REFERENCES

Bertrand, J. W. M., & Fransoo, J. C. (2002). Operations management research methodologies using quantitative modeling. *International Journal of Operations and Production Management*, 22(2), 241–264. https://doi.org/10.1108/01443570210414338

Caiado, R. G. G., Scavarda, L. F., Gavião, L. O., Ivson, P., Nascimento, D. L. de M., & Garza-Reyes, J. A. (2021). A fuzzy rule-based Industry 4.0 maturity model for operations and supply chain management. *International Journal of Production Economics*, 231, 107883. https://doi.org/10.1016/j.ijpe.2020.107883

Cebollada, S., Payá, L., Flores, M., Peidró, A., & Reinoso, O. (2020). A state-of-the-art review on mobile robotics tasks using artificial intelligence and visual data. *Expert Systems with Applications*, June, 167, 114195. https://doi.org/10.1016/j.eswa.2020.114195

Collins, C., Dennehy, D., Conboy, K., & Mikalef, P. (2021). Artificial intelligence in information systems research: A systematic literature review and research agenda. *International Journal of Information Management*, 60, 102383. https://doi.org/10.1016/j.ijinfomgt.2021.102383

del Mar Roldán-García, M., García-Nieto, J., & Aldana-Montes, J. F. (2018). Enhancing semantic consistency in anti-fraud rule-based expert systems. *Actas de Las 23rd Jornadas de Ingenieria Del Software y Bases de Datos, JISBD 2018*, 332–343. https://doi.org/10.1016/j.eswa.2017.08.036

De Oliveira, E. M., Leme, D. S., Barbosa, B. H. G., Rodarte, M. P., & Alvarenga Pereira, R. G. F. (2016). A computer vision system for coffee beans classification based on computational intelligence techniques. *Journal of Food Engineering*, *171*, 22–27. https://doi.org/10.1016/j.jfoodeng.2015.10.009

Dulebenets, M. A. (2019). A delayed start parallel evolutionary algorithm for just-in-time truck scheduling at a cross-docking facility. *International Journal of Production Economics*, *212*, 236–258. https://doi.org/10.1016/j.ijpe.2019.02.017

Elshaer, R., & Awad, H. (2020). A taxonomic review of metaheuristic algorithms for solving the vehicle routing problem and its variants. *Computers and Industrial Engineering*, *140*, 106242. https://doi.org/10.1016/j.cie.2019.106242

Farhan, W., Talafha, B., Abuammar, A., Jaikat, R., Al-Ayyoub, M., Tarakji, A. B., & Toma, A. (2020). Unsupervised dialectal neural machine translation. *Information Processing and Management*, *57*(3), 102181. https://doi.org/10.1016/j.ipm.2019.102181

Hosseini, S., & Ivanov, D. (2020). Bayesian networks for supply chain risk, resilience and ripple effect analysis: A literature review. *Expert Systems with Applications*, *161*, 113649. https://doi.org/10.1016/j.eswa.2020.113649

IFORS. (2022). *Definition of Operations Research*. International Federation of Operational Research Societies. Retrieved from https://www.ifors.org/what-is-or/

Kang, Z., Catal, C., & Tekinerdogan, B. (2020). Machine learning applications in production lines: A systematic literature review. *Computers and Industrial Engineering*, *149*, 106773. https://doi.org/10.1016/j.cie.2020.106773

Kim, M. C., Kim, C. O., Hong, S. R., & Kwon, I. H. (2008). Forward-backward analysis of RFID-enabled supply chain using fuzzy cognitive map and genetic algorithm. *Expert Systems with Applications*, *35*(3), 1166–1176. https://doi.org/10.1016/j.eswa.2007.08.015

Klaib, A. F., Alsrehin, N. O., Melhem, W. Y., Bashtawi, H. O., & Magableh, A. A. (2021). Eye tracking algorithms, techniques, tools, and applications with an emphasis on machine learning and Internet of Things technologies. *Expert Systems with Applications*, *166*, 114037. https://doi.org/10.1016/j.eswa.2020.114037

Lee, W. K., Leong, C. F., Lai, W. K., Leow, L. K., & Yap, T. H. (2018). ArchCam: Real time expert system for suspicious behaviour detection in ATM site. *Expert Systems with Applications*, *109*, 12–24. https://doi.org/10.1016/j.eswa.2018.05.014

Lima-Junior, F. R., & Carpinetti, L. C. R. (2019). Predicting supply chain performance based on SCOR ® metrics and multilayer perceptron neural networks. *International Journal of Production Economics*, *212*, 19–38. https://doi.org/10.1016/j.ijpe.2019.02.001

Mitroff, I. I., Betz, F., Pondy, L. R., & Sagasti, F. (1974). On managing science in the systems age: Two schemas for the study of science as a whole systems phenomenon. *Interfaces*, *4*(3), 46–58. https://doi.org/10.1287/inte.4.3.46

Paschen, J., Kietzmann, J., & Kietzmann, T. C. (2019). Artificial intelligence (AI) and its implications for market knowledge in B2B marketing. *Journal of Business and Industrial Marketing*, *34*(7), 1410–1419. https://doi.org/10.1108/JBIM-10-2018-0295

Sabahi, S., & Parast, M. M. (2020). The impact of entrepreneurship orientation on project performance: A machine learning approach. *International Journal of Production Economics*, *226*, 107621. https://doi.org/10.1016/j.ijpe.2020.107621

Silva, A., Aloise, D., Coelho, L. C., & Rocha, C. (2021). Heuristics for the dynamic facility location problem with modular capacities. *European Journal of Operational Research*, *290*(2), 435–452. https://doi.org/10.1016/j.ejor.2020.08.018

Smiti, S., & Soui, M. (2020). Bankruptcy prediction using deep learning approach based on borderline SMOTE. *Information Systems Frontiers*, *22*(5), 1067–1083. https://doi.org/10.1007/s10796-020-10031-6

Sun, T. Q., & Medaglia, R. (2019). Mapping the challenges of artificial intelligence in the public sector: Evidence from public healthcare. *Government Information Quarterly*, *36*(2), 368–383. https://doi.org/10.1016/j.giq.2018.09.008

Syed, R., Suriadi, S., Adams, M., Bandara, W., Leemans, S. J. J., Ouyang, C., ter Hofstede, A. H. M., van de Weerd, I., Wynn, M. T., & Reijers, H. A. (2020). Robotic process automation: Contemporary themes and challenges. *Computers in Industry*, *115*. https://doi.org/10.1016/j.compind.2019.103162

Tellez, E. S., Miranda-Jiménez, S., Graff, M., Moctezuma, D., Siordia, O. S., & Villaseñor, E. A. (2017). A case study of Spanish text transformations for twitter sentiment analysis. *Expert Systems with Applications*, *81*, 457–471. https://doi.org/10.1016/j.eswa.2017.03.071

Torres-Jiménez, J., & Pavón, J. (2014). Applications of metaheuristics in real-life problems. *Progress in Artificial Intelligence*, *2*(4), 175–176. https://doi.org/10.1007/s13748-014-0051-8

Wang, H., Ding, S., Wu, D., Zhang, Y., & Yang, S. (2019). Smart connected electronic gastroscope system for gastric cancer screening using multi-column convolutional neural networks. *International Journal of Production Research*, *57*(21), 6795–6806. https://doi.org/10.1080/00207543.2018.1464232

Uncertainty and Improving Operations Management in the Age of Digital Transformation

10

DOI: 10.1201/9781003226468-10

10.1 UNCERTAINTY AND ENVIRONMENTAL COMPLEXITIES

Nowadays, organizations and their supply chains are operating in an environment in which uncertainties in their business activities and operations have increased drastically. This chapter explores uncertainty and environmental complexities from an integrated perspective. In other words, environmental events can significantly amplify uncertainty levels, thus imposing several challenges upon organizations.

Furthermore, the contemporary world is experimenting with unprecedented, intertwined, simultaneous, and prolonged disruptions (Fosso Wamba et al., 2022; Ivanov & Dolgui, 2020), which in turn affect (main negatively, although sometimes positively) all types of organizations. Complex and interconnected events such as the COVID-19 pandemic, the Russia–Ukraine war, climate changes, financial crises, and terrorism can generate lasting side effects for organizations and consequently for society.

With the digital transformation, companies can face these and other types of emergency situations by employing cutting-edge technologies in their operations and supply chains (Fosso Wamba et al., 2021). In this context, it is clear that digital transformation can minimize the negative effects of severe environmental disruptions provoked by complex events (Ye et al., 2022).

10.2 A DIGITAL TRANSFORMATION FRAMEWORK TO OVERCOME ENVIRONMENTAL UNCERTAINTIES AND DISRUPTIONS

In order to provide a holistic view of the digital transformation, its most representative technologies, and interaction with humans and organizations, while also taking environmental uncertainties into account, this section provides a novel framework to support companies, decision-makers, and scholars, and provide an in-depth understanding of how the digital transformation phenomenon can contribute to minimizing the negative effects of environmental uncertainties and their resulting disruptions.

Figure 10.1 presents a framework to support uncertainties as well as environmental disruptions with digital transformation approaches. The framework has five dimensions. The first two are associated with uncertainties and environmental disruptions (first, simultaneous and intertwined challenges, and second, negative effects). The third dimension is the application of the TOP framework, which in turn enables the last two dimensions, benefits to operations and supply chains and general benefits.

10.2.1 Simultaneous and Intertwined Challenges

In today's industry, complex challenges can interact with each other and consequently bring about unprecedented consequences. Uncertainties and environmental disruptions can amplify the negative effects and consequences of a set of challenges such as war risks, humanitarian crises, pandemic outbreaks, climate changes, financial crises, fake news, political issues, or a lack of trained human resources. All or part of these complexities/events can interact with each other and amplify the negative effects over the long term.

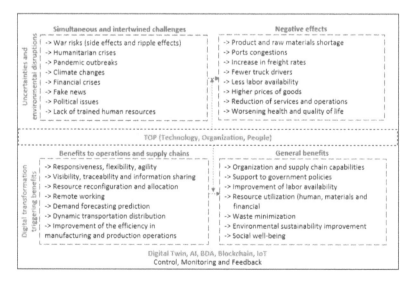

FIGURE 10.1   Tackling uncertainties and environmental disruptions with digital transformation.

10.2.2 Negative Effects

Simultaneous and intertwined challenges can produce concurrent negative effects. For instance, product and raw materials shortages, port congestion, increased freight rates, fewer truck drivers, labor shortages, higher prices of goods, and the reduction of services and operations can lead to a worsening of health and quality of life. As highlighted previously, the duration of the negative effects is unknown; they may appear not only in the short term but also in the middle and long terms.

10.2.3 The Role of the TOP Framework

In order to address the previous two dimensions (simultaneous and inter-twined challenges and negative effects), the framework considers the utilization of this book's core TOP framework.

From the technology side (see Chapter 2), to minimize the negative effects while also creating the resilience and adaptability of organizations and supply chains, the use of cutting-edge technologies, like the Internet of Things, big data analytics, artificial intelligence, simulation, additive manu-facturing, blockchain, and digital twin, can leverage predictive power and responsiveness.

With regard to the company itself, an organizational culture that dis-seminates the importance of and need for a continuous digital transformation is essential to support managers' actions, workers, different departments, and supply chains. Furthermore, a limited digital culture implies less top manage-ment support for reactions and anticipatory actions. As a result, the negative effects can be amplified and have a longer impact.

Considering the people dimension of the TOP framework, the conver-gence and operationalization of an organization's plans require clear commu-nication between departments and workers. Moreover, workers with strong skills—not only in technologies but also in communication, leadership, and strategies—are fundamental to overcome negative effects.

10.2.4 Benefits to Operations and Supply Chains

The digital transformation, represented here by the TOP framework, works as a trigger in creating and or leveraging benefits. Some of the main benefits to the operations and supply chains include responsiveness, flexibility, agility, visibility, traceability and information sharing, resource reconfiguration and allocation, remote working, demand forecasting, dynamic transportation distribution, and improvement of the efficiency in manufacturing and production operations. These benefits can improve the resilience of not only an organization but the entire supply chain.

10.2.5 General Benefits

The general benefits dimension reflects the final part of the framework, which is focused mainly on the stakeholder's perspective. Accordingly, the following general benefits can emerge with the digital transformation applied to uncertainties and environmental disruptions: organization and supply chain capabilities, support for government policies, improvement of labor availability, resource utilization (human, materials, and financial), waste minimization, environmental sustainability improvement, and social well-being.

10.3 SHORT CASE STUDY: MINIMIZING THE NEGATIVE EFFECTS OF AN AGRICULTURE SUPPLY CHAIN EXPORTER

Company XKO is one of the largest soybean producers in Latin America. Its exports have met worldwide demand. But the company has been facing several challenges in recent months, including the COVID-19 pandemic, climate changes, increased freight rates, and difficulties booking freight due to the scarcity of containers. Year after year, the company has increased its productivity; however, revenues are stagnant and, in some countries, have decreased. The directors of XKO are focused on the operating business model and efforts in farm production. A survey of XKO workers showed that the XKO operations and supply chains are basically managed by spreadsheets, and there is no formal flow for how goods should be exported. The company makes shipments according to customer orders. Using the framework from Figure 10.1, show how XKO can minimize these negative effects.

REFERENCES

Fosso Wamba, S., Queiroz, M. M., Roscoe, S., Phillips, W., Kapletia, D., & Azadegan, A. (2021). Guest editorialEmerging technologies in emergency situations. *International Journal of Operations & Production Management, 41*(9), 1405–1416. https://doi.org/10.1108/IJOPM-09-2021-904

Fosso Wamba, S., Queiroz, M. M., & van Hoek, R. (2022). *Call for Proposals Special Topic Forum: Disruptive-Techs and the (Real) Value Creation to Firms and Supply Chains in Today's Industry: A Proactive Perspective.* Journal of Business Logistics. https://onlinelibrary.wiley.com/pb-assets/assets/21581592/STF%20-%20Does%20Disruptive%20Tech%20Matter[78]-1647962591183.pdf

Ivanov, D., & Dolgui, A. (2020). Viability of intertwined supply networks: Extending the supply chain resilience angles towards survivability. A position paper motivated by COVID-19 outbreak. *International Journal of Production Research, 58*(10), 2904–2915. https://doi.org/10.1080/00207543.2020.1750727

Ye, F., Liu, K., Li, L., Lai, K.-H., Zhan, Y., & Kumar, A. (2022). Digital supply chain management in the COVID-19 crisis: An asset orchestration perspective. *International Journal of Production Economics, 245*, 108396. https://doi.org/10.1016/j.ijpe.2021.108396

Milton Keynes UK
Ingram Content Group UK Ltd.
UKHW020626121223
434203UK00004B/30